Palgrave Studies in Impact Finance

Series Editor
Mario La Torre
Sapienza University of Rome
Rome, Italy

The *Palgrave Studies in Impact Finance* series provides a valuable scientific 'hub' for researchers, professionals and policy makers involved in Impact finance and related topics. It includes studies in the social, political, environmental and ethical impact of finance, exploring all aspects of impact finance and socially responsible investment, including policy issues, financial instruments, markets and clients, standards, regulations and financial management, with a particular focus on impact investments and microfinance.

Titles feature the most recent empirical analysis with a theoretical approach, including up to date and innovative studies that cover issues which impact finance and society globally.

More information about this series at
http://www.palgrave.com/gp/series/14621

Marco Migliorelli • Philippe Dessertine
Editors

The Rise of Green Finance in Europe

Opportunities and Challenges for Issuers, Investors and Marketplaces

palgrave
macmillan

Editors
Marco Migliorelli
IAE Université Paris 1 Panthéon-Sorbonne (Sorbonne Business School)
Paris, France

European Commission, Brussels
Belgium

Philippe Dessertine
IAE Université Paris 1 Panthéon-Sorbonne (Sorbonne Business School)
Paris, France

ISSN 2662-5105 ISSN 2662-5113 (electronic)
Palgrave Studies in Impact Finance
ISBN 978-3-030-22509-4 ISBN 978-3-030-22510-0 (eBook)
https://doi.org/10.1007/978-3-030-22510-0

This Palgrave Macmillan imprint is published by the registered company Springer Nature Switzerland AG.
The registered company address is: Gewerbestrasse 11, 6330 Cham, Switzerland

Preface

The issue of the sustainability of human activities is increasingly debated at political and societal level, in particular, following the adoption of the Sustainable Development Goals (SDGs) in the context of the United Nations Development Programme (UNDP) in September 2015 and the Paris Agreement reached in the United Nations Framework Convention on Climate Change (UNFCCC) in December of the same year. In 2018, the Intergovernmental Panel on Climate Change (IPCC)—the United Nations body responsible for assessing the science related to climate change—further launched the alarm stating that the world needs to limit temperature increase to 1.5°C with respect to pre-industrial levels to reduce the likelihood of extreme weather events and emphasized that greenhouse gases emissions need to be reduced with far more urgency than previously assumed. Net-zero carbon emissions at the global level need to be achieved not beyond the half of this century and neutrality for all other greenhouse gases not much later.

The discussion on how to lead the sustainability transition embraces different disciplines. Aspects regarding political engagement, effective regulation, technological improvements, scientific research and investment flows will jointly determine the feasibility and the speed of this changeover. In this respect, the relevance of finance as an essential enabling factor is constantly increasing. *Green finance*, as a crucial component of the sustainable finance landscape, has emerged only in the last decade. Nevertheless, today it is commonly considered one of the most promising contributors to the achievement of the environmental goals. From the launch of the first green bond by the European Investment Bank in 2007, the market of green securities has gained substantial momentum and has reached a record level of global green bond issuance of USD 155.5 billion in 2017. Sustained by a strong demand (in particular from institutional investors), the market of green securities is indeed quickly evolving. On the one hand, large corporations and banks strive to be recognised as environmentally responsible. To this extent, they seek to obtain green labels for their debt issuances aimed at financing green projects (independent certification bodies have emerged for this purpose) and have started to disclose information on their sustainability-related activities (even though no industry standards still exist on the quality and quantity of information to disclose). On the other hand, stock exchanges and fund managers have started to adapt their offering. In order to sustain the growth of the market and seize a growing business opportunity, dedicated listing options and a wide range of green investment solutions are progressively getting available. Finally, governments and public administrations are also gradually entering the market. For these actors, the issuance of green securities is an effective means to link their political commitments to the consequent financing needs. However, a number of issues that can potentially harm the credibility of the nascent market remain unresolved. The reliability of existing green labels, the possible lack of control on the use of proceeds, the threat of green washing and the uncertainty of an effective economic return for issuers of green securities are some of these issues. In addition, it seems increasingly evident that market dynamics will hardly be sufficient to trigger a sufficient amount of financial resources to reach the most ambitious environmental objectives, and some form of policy interventions is indeed necessary.

In this scenario, it is of the utmost interest for practitioners, academicians, policy makers and the general public to have a clear understanding of the key features of the *green finance* market as it stands today. To this extent, this book, which is in the form of an edited collection, offers a comprehensive discussion of how this sector has been growing so far and on what are the opportunities and challenges ahead by including some reflections on the main hurdles today to mainstream *green finance*. The specific focus of the book is on Europe in order to emphasise the policy actions undertaken in the continent and their impacts on the development of the market. Nevertheless, as many aspects of *green finance* have worldwide scope and implications, the discussion is widened whenever necessary.

The book is structured in two parts. Part I, which focuses on the analysis of state of the art, comprises six chapters. In Chap. 1, Romain Berrou,

Philippe Dessertine and Marco Migliorelli present an overview of the key characteristics of the *green finance* market. To this extent, the chapter first recalls the political processes culminating with the Paris Agreement and the adoption of the Sustainable Development Goals, as well as highlights the role of *green finance* in these processes. Hence, it provides a detailed picture of the market by describing the types of green securities and showing the recent investment trends. In Chap. 2, Romain Berrou, Nicola Ciampoli and Vladimiro Marini provide an assessment of the issues linked to the lack of a clear definition of *green finance*. In this respect, they review the main approaches in use today for determining which sectors are eligible for "green" funding, assess the main principles adopted by the industry to label a financial security as "green" and discuss the risk of greenwashing. Chapter 3 by Olaf Weber and Amr ElAlfy analyses the development of *green finance* by sector. To this extent, it examines the role of multilateral development banks and the approaches adopted by industrial companies and financial institutions. Chapter 4, authored by Giovanni Ferri and Francesca Lipari is dedicated to the discussion on the management of sustainable finance within the company. The authors first present the different challenges ahead while introducing measures to foster sustainable and green finance initiatives. Then, they provide some empirical check of the nexus between organisational structures and environmental, social and governance (ESG) ratings. In Chap. 5, Dirk Schiereck, Gunnar Friede and Alexander Bassen give an overview of the issues linked to the performances of green securities. To do that, they analyse the state of the literature on the performance of the green bonds as well as on the relationship between ESG performances and financial performances of the company. Finally, Chap. 6, written by Vladimiro Marini, presents the major initiatives undertaken in the European Union in the last decade as concerns the sustainable finance sector. In this respect, he first highlights the key role of the European Investment Bank in triggering the development of the market and then in becoming one of the most important issuers of green securities in the world. Hence, he summarizes the main actions launched by the European Commission, including the recent plans and legislative proposals to support sustainable finance.

Part II of the book is devoted to the analysis of the (long) way forward and the new opportunities ahead for *green finance*, and it is composed of five chapters. In Chap. 7, Marco Migliorelli and Philippe Dessertine analyse to what extent market-based dynamics can still support the development of *green finance* and what are the expected limits of such a transaction-based model. Then, they focus their reasoning on the main areas of intervention in order to mainstream *green finance* by pointing out in particular the importance of coherently factoring-in environmental risks in the investors' decision-making process, of encouraging the banking sector to embrace *green finance* and of pushing policy makers to structurally support the sustainable finance movement. In Chap. 8, Marco Migliorelli focuses his analysis on the causes of the limited role of *green finance* in agriculture (which is responsible for about 25% of the greenhouse gas emission), with specific reference to the case of the European Union. He first identifies the main obstacles that still today hinder the development of green financing in this sector. Hence, he discusses possible ad hoc financing structures that can contribute to reverse the observed tendency. In Chap. 9, Gregor Dorfleitner and Diana Braun examine the potential of fintech and blockchain to unlock the mobilisation of *green finance* and to overcome respective barriers. To this extent, they explain the key functionalities of possible concrete applications, including their key benefits and limitations. Chapter 10, written by Silvio Goglio and Ivana Catturani, provides a reflection on the potential impact of sustainable finance on the European social model of the future. To do that, they discuss the need to fully consider the positive externalities produced by sustainable investments and the pivotal role of financial intermediaries in driving resources towards activities producing green and social capital. Finally, in Chap. 11, Marco Migliorelli and Philippe Dessertine summarise the main issues and trends discussed in the book and present some concluding remarks.

Paris, France and Brussels, Belgium — Marco Migliorelli
Paris, France — Philippe Dessertine

Acknowledgments

We would like to thank the contributors. This book would have not come to fruition without their strong expertise and exceptional commitment.

Contents

Notes on Contributors

Alexander Bassen is a full professor at the University of Hamburg, where he also holds the Chair for Capital Markets and Management. He holds a PhD from the European Business School of Oestrich-Winkel. His research primarily focuses on environment, social and governance matters in the context of capital markets and management.

Romain Berrou is a PhD candidate in finance at the IAE of the University Paris 1 Panthéon-Sorbonne (Sorbonne Business School). His research interests include green and sustainable finance and financing instruments innovation in the agricultural sector.

Diana Braun is a researcher and doctoral student at the University of Regensburg. She holds a Master of Science degree in economics from the University of Munich. Her research focus is on green and sustainable finance. She has professional experience as a strategic consultant in the financial service sector.

Ivana Catturani is a research fellow and lecturer at the University of Trento and a researcher at Euricse. She obtained her PhD in economics and management at the University of Trento. Her research activity is focused on cooperative financial institution, with a particular focus on their governance issues and their relevance for local development.

Nicola Ciampoli is an independent advisor, an accountant and a lecturer of financial intermediaries at the LUMSA University in Rome. He holds a PhD from the University of Rome Tor Vergata. His research interests include sustainable value in banking and banking regulation.

Philippe Dessertine is a full professor of finance at the IAE of the University Paris 1 Panthéon-Sorbonne (Sorbonne Business School), the director of *Institut de Haute Finance*, Paris, France, and a former member of the *Haut Conseil des finances publiques* in France. He is the author of several publications on the role of finance in the modern society.

Gregor Dorfleitner is a full professor of finance at the University of Regensburg and the director of the Center of Finance at the same university. He is an associate researcher at the Centre of European Research in Microfinance (CERMi) in Belgium. He is the author of more than 60 international publications on diverse topics, including fintech, crowdfunding, socially responsible investing and microfinance.

Amr ElAlfy is a PhD candidate and sessional instructor at the School of Environment, Enterprise and Development (SEED), the University of Waterloo, Canada. His research focuses on corporate social responsibility, impact measurement and capacity development. He has professional experience in the marketing function in multinational companies.

Giovanni Ferri is a full professor at the LUMSA University in Rome, where he is also the director of the Master of Science in Management of Sustainable Development Goals. He worked at the World Bank and Bank of Italy, handling apical research and policy roles. His research interests include financial market imperfections, the transmission of monetary and financial shocks, corporate governance and sustainability.

Gunnar Friede is a senior fund manager and director at Deutsche Asset Management (formerly DWS Investments), where he focuses on the firm's global environmental, social and governance (ESG) strategy. He is a Certified European Financial Analyst (CEFA) and Certified International Investment Analyst (CIIA). He is the author of several publications on the relation between corporate financial and ESG performances.

Silvio Goglio is an associate professor of political economy at the University of Trento and coordinator of the research area in cooperatives and development at Euricse. His research interests include collective efficiency and local governance, institutional changes and economic performance, local and cooperative credit and regional development.

Francesca Lipari is a research fellow at the LUMSA University in Rome. Previously, she was a research fellow at the University of Pennsylvania. She holds a PhD from the University of Rome Tor Vergata. Her research interests include sustainability, behavioral economics, network theory and political economy of development.

Vladimiro Marini is a PhD candidate in banking and finance at the University of Rome Tor Vergata and a junior assistant professor at Roma Tre University. His research activity is primarily focused on private equity investments, in particular, as concerns the skills and competences acquisition by fund managers.

Marco Migliorelli is a researcher in finance at the IAE of the University Paris 1 Panthéon-Sorbonne (Sorbonne Business School), and an economist at the European Commission. He holds a PhD from the University of Rome Tor Vergata. His research interests include green finance, cooperative banking and financial instruments innovation.

Dirk Schiereck is s a full professor of corporate finance at the Darmstadt University of Technology. He is a member of the supervisory board of the exchange-listed German fintech start-up Creditshelf. He holds a PhD from the University of Mannheim and he is the author of about 300 international publications on a broad range of finance topics, including energy finance, empirical corporate finance and bond investing.

Olaf Weber is a full professor at the University of Waterloo, where he also holds the Chair in Sustainable Finance. He is the editor-in-chief of the Journal of Sustainable Finance and Investment. He holds a PhD from the University of Bielefeld. His research interests lie in the connection between financial sector players and sustainable development, and the link between sustainability and financial performance of enterprises.

List of Figures

List of Tables

PART I

State of the Art

CHAPTER 1

An Overview of Green Finance

Romain Berrou, Philippe Dessertine, and Marco Migliorelli

1.1 Introduction

The modern world's economic system has led to a global environmental crisis (e.g. IPCC 2018). As this view is slowly starting to be embraced by businesses and financial institutions, financial markets are evolving to provide new forms of funding to actors that wish to face this crisis. Even if always bearing in mind its risk-return priorities, part of the financial market is nowadays joining public actors, non-governmental organisations (NGOs) and

The contents included in this chapter do not necessarily reflect the official opinion of the European Commission. Responsibility for the information and views expressed lies entirely with the authors.

R. Berrou • P. Dessertine (✉)
IAE Université Paris 1 Panthéon-Sorbonne (Sorbonne Business School), Paris, France
e-mail: Romain.Berrou@etu.univ-paris1.fr; philippe.dessertine@univ-paris1.fr

M. Migliorelli (✉)
IAE Université Paris 1 Panthéon-Sorbonne (Sorbonne Business School), Paris, France

European Commission, Brussels, Belgium

M. Migliorelli, P. Dessertine (eds.), *The Rise of Green Finance in Europe*, Palgrave Studies in Impact Finance,
https://doi.org/10.1007/978-3-030-22510-0_1

civil society in the efforts to face the global environmental challenge. The funding of these efforts is what can be broadly referred to as *green finance*.[1]

As the portion of the financial markets that focuses on solving the environmental crisis grows larger and more diverse, it carries with it the notion that capital can be used to solve extra-financial issues in addition to providing funding and generating profits. Evidently, this notion is not new,[2] and some might even say it has existed since the first barter-like methods of exchange. In comparison, both capitalism and modern finance are relatively young, the former becoming the encompassing global economic system only in the late twentieth century and the latter roughly considered to be born in the 1950s with the first seminal works on modern financial theory (e.g. Markowitz 1952). Nevertheless, the concept that natural resources are limited and that some by-products of mass production processes (e.g. polluting gases or waste) are highly detrimental to both humankind and its natural environment was not central in the public debate until recent years. Capitalism and modern finance could indeed be considered as an effective system and an efficient means to trigger the economic potential of nations irrespective of their impact on the environment. As a matter of fact, the twentieth century also corresponds to a period of great economic growth and of exceptional improvement in human conditions in those countries that started their industrialisation processes. Global warming and the consequent increasing incidence of climate-related extreme weather events (such as droughts, floods and storms)[3] have then worked as a wake-up call on the limits of any economic model that does not foresee the preservation of the environment as one of its pillars. In this respect, *green finance* represents the global financial community's first structured attempt to join financial performances and positive environmental impact, and can be seen as one of the concrete signs of the economic system's adaptation to the global environmental challenge.

[1] As of today, a unique and well-established definition of *green finance* does not exist. Several definitions have been proposed by the industry and by international organisations for specific green securities and products as well as for the different components of the sustainable finance industry. A more detailed dissertation about the definitions used in the field of *green finance* is given in Chap. 2.

[2] An example is given by cooperative and savings financial institutions, which formally adopted since the nineteenth century in Europe a dual bottom-line approach for which financial performance needed to coexist with social goals (such as mutuality, financial and social inclusion, support to the local communities). For more details, see for example Ayadi et al. (2010) or Migliorelli (2018).

[3] For a thorough dissertation of the detrimental consequences of an increase in the global temperatures, see again IPCC (2018).

In such a context, this chapter provides an overview of the main developments in *green finance* from its origins to the present. In this respect, the main roots of the role of ethics in finance are first recalled. Then, the major international events contributing to raise the attention to the need to preserve the environment are treated, in particular as concerns the *Paris Agreement* and the adoption of the *Sustainable Development Goals* (SDG). Hence, a discussion is provided on the role of *green finance* today within the sustainable finance landscape, including an outline on the main families of existing financial products and services and their evolution over time. Finally, the key challenges still ahead for *green finance* in order to be considered a stable component of the modern financial landscape are mentioned.

1.2 Ethics and Sustainability in Finance

Signs of humankind's concern with the notion of ethical use of money can already be found in religious texts. The ***usurer***, that is, a lender compensated through (unjustified) interests on the money lent, was, for example, already mentioned in the *Vedas*, a body of writings from ancient India considered as the oldest scriptures of Hinduism, drafted between 1700 and 1100 BC. The notion of usury is then present in the religious texts of Buddhism, Judaism, Christianity and Islam.[4] Many movements developed over time on similar basis,[5] and some argue that the attitude towards the use of money as influenced by religious beliefs (and hence ethics) can be interpreted as the early form of what can be considered today as ***sustainable finance***.[6]

[4] Even today, many Muslims, on the basis of the precepts of *Sharia*, avoid practices of lending that result in any kind of *riba* (interest) and investing in industries such as tobacco, alcohol or gambling. These are some of the elements that feature in the modern *Islamic finance*. For a more accurate analysis, see for example Ayub (2013) or Warde (2000).

[5] For example, in England, a law called *The Act Against Usury*, which prohibited excessive interests on loans was in effect from 1571 to 1624 (Glaeser and Scheinkman 1998; Lewison 1999). In the mid-1750s, Quakers banned their followers from participating in the slave trade, and believers in the Methodist Church were asked to avoid investments in weapons, tobacco, alcohol or gambling, or any trade that could be harmful to life and health (Renneboog et al. 2008).

[6] *Sustainable finance* can be broadly defined as the stocks and flows of financial resources and assets (across banking, investment and insurance industries) which is aligned with a large range of environmental, social and economic objectives and more generally with the delivery of the *Sustainable Development Goals* (SDGs) as developed in the context of the *United*

However, the relationship between money and ethics had a strong acceleration only in modern times and eventually became not restricted to the boundaries of religious communities. Instances of this change can be found in the follow-up of social protests that occurred in the United States in the 1950s and 1960s, through both the civil rights movement and the increasing opposition to the involvement in the Vietnam War. For example, movements that opposed the use of *agent orange* (a highly toxic gas developed to destroy forests and which frightened indigenous populations during the Vietnam War) led to the creation of the *Pax World Balanced Fund* in 1971, a fund that was explicitly thought for investors that wished to avoid direct investments in any firm that participated in the production of this gas. As a matter of fact, the *Pax World Balanced Fund* can be considered one of the first examples of socially responsible funds. In this respect, it was closely followed by the creation of the *Dreyfus Third Century Fund* in 1972, which actively looked to invest in best-performing firms in terms of the enhancement of quality of life in the United States. As these investment strategies grew in popularity and appeal, initiatives in which finance was used as a tool to enforce social justice proliferated. One of the most compelling examples of large-scale initiatives was the drafting of the *Sullivan principles*[7] in 1977, a code of conduct for businesses that factually contributed to the movement that determined the end of *apartheid* in South Africa.

During the last decades of the twentieth century, the socially responsible investing movement grew further (e.g. Renneboog et al. 2008), benefitting from three main developments, which finally created the basis for an increasingly severe assessment by investors as concerns social and environmental matters. The first referred to the multiplying warnings by scien-

Nation Development Program (UNDP). In this respect, *green finance* should be considered a fundamental component of sustainable finance. For more details, see also UNEP (2016).

[7] The original *Sullivan principles* were developed by the African-American preacher Rev. Leon Sullivan, promoting corporate social responsibility and to apply economic pressure on South Africa in protest of its system of *apartheid*. The principles eventually gained wide adoption among US-based corporations. As investors realised that some American firms did not make any particular efforts to end discrimination in South Africa, these firms were eventually hit by a massive wave of divestment. The success of this initiative led in 1999 to a new global version of the *Sullivan principles*. Jointly unveiled by Rev. Sullivan and United Nations Secretary General Kofi Annan, the new and expanded code of conduct, as opposed to the originals' specific focus on South African *apartheid*, was designed to increase the active participation of corporations in the advancement of human rights and social justice at the international level.

tists and researchers regarding the dangers of climate change. The second concerned the increasing acceptance of the view that poor corporate governance can be detrimental to both markets and the firm's impact on the society and the environment. The third consisted in the widening of the debate on the relationship between institutional investors' fiduciary duty and sustainability issues. As a result, those that can today be referred to as socially responsible investors started at that time to consider the deterioration of the environment as a specific cause for concern and to increasingly apprise conscientious firms in terms of environmental management. In particular following the *Three Mile Island* and then the *Chernobyl* nuclear power plant accidents, occurring in 1979 and 1986, respectively, and the *Exxon Valdez* shipwreck in 1989 (spilling of about 41,000 cubic metres of crude oil near the coasts of Alaska), global awareness regarding the possible consequences of bad business practices on the environment reached an unprecedented peak. In this regard, 1989 was also the year in which the first official *Socially Responsible Investing* (SRI) conference was held.[8]

In parallel with civil society and private initiatives, also the global community of nations gradually embraced the sustainability movement. In 1972 the *United Nations Conference on the Human Environment* led to the creation of the *United Nations Environmental Programme* (UNEP). Particularly on topics of fiduciary duty and sustainability, the UNEP started its work on the link between *environmental, social and governance* (ESG) factors and financial performance (even though a report in which the organisation concluded that "integrating ESG considerations into an investment analysis [...] is clearly permissible and is arguably required in all jurisdictions" was published only many years later, in 2005[9]). In 1988, UNEP and the *World Meteorological Organization* (WMO) created the *Intergovernmental Panel on Climate Change* (IPCC), a body charged with providing a scientific view on climate change and its impact on society. Next year, the *Coalition for Environmentally Responsible Economics*

[8] Defining precisely what constitutes the SRI industry may be a challenge. One possible current definition is given by Eurosif, a leading European association for the promotion and advancement of the SRI industry in Europe. Based on this definition, "*Sustainable and responsible investment* (SRI) is a long-term oriented investment approach which integrates environmental, societal and governance (ESG) factors in the research, analysis and selection process of securities within an investment portfolio. It combines fundamental analysis and engagement with an evaluation of ESG factors in order to better capture long term returns for investors, and to benefit society by influencing the behaviour of companies".

[9] See UNEP (2005).

(CERES)[10] was established, which in turn laid the foundation for the UNEP's *Global Reporting Initiative* (GRI), which enhanced global focus on environmental transparency for large firms.

1.3 COP 21 and the Paris Agreement

1.3.1 Main Steps Leading to COP 21

The *United Nations Conference on the Human Environment*, which created the UNEP in 1972, was not immediately successful, and the global community did not organise a similar event for another 20 years. This can be in part explained by the fact that at that time the United Nations (UN) could not benefit from a real global influence due to the ongoing Cold War and that most of its resources were indeed oriented towards its primary peacekeeping missions.[11] The *United Nations Conference on Environment and Development* (also known as *Rio Summit* or *Earth Summit*) took place only in 1992. This event gave birth to many initiatives that are still central to most global debates addressing climate change. It first led to the development of *Agenda 21*, an action plan for the UN, multilateral organisations and governments that aimed to achieve global sustainable development through local, national and global action. It then resulted in the *United Nations Framework Convention on Climate Change* (UNFCCC), an international environmental treaty that aimed to limit global greenhouse gas emissions and that is still in force today. Starting from 1995, signatories of the UNFCCC met on a yearly basis, through the *Conferences of the Parties* (COP). In 1997, as a result of the conference held in Kyoto (COP 3), the *Kyoto Protocol* extended on the UNFCCC and

[10] The CERES is a non-profit organisation based in the United States which comprises investors and environmental, religious and public interest groups. The organisation's purpose is to promote investment policies that are environmentally, socially and financially sound. There are currently more than 70 members, including large financial organisations.

[11] The primary mission of UN, after peacekeeping, security and human rights, is economic development and humanitarian assistance. These objectives led to the creation of well-known international organisations that focused on some of its specific underlying missions. This included global international development (*United Nations Development Program*, UNDP), agricultural development and food security (*The Food and Agriculture Organization*, FAO), aid for children around the world (*United Nations Children's Fund*, UNICEF), funding of international development and global access to finance (*World Bank Group*) and global economic cooperation (*International Monetary Fund*, IMF) amongst others.

led to the establishment of the first global legally binding obligation addressing climate change.[12]

The period that extends from the 1997 conference in Kyoto to the 2015 *United Nations Climate Change Conference* (COP 21) in Paris was indeed not one of great progress for global negotiations on climate change. In fact, the Kyoto Protocol entered into force only in 2005, and it did so without the participation of the United States.[13] A few notable events took place only in the period close to COP 21, which could provide indications of a shift in attitudes. In 2014, the European Union (EU) committed to cutting its greenhouse gas emissions by 40% by 2030 with respect to the 1990 levels, a few months before the United States and China drew up a similar arrangement. That same year, the *People's Climate March* led over 300.000 participants to protest in the streets of New York on the subject of climate change. In May 2015, Pope Francis addressed the subject of environmental degradation and climate change in a historical encyclical.[14]

1.3.2 The Conference and the Paris Agreement

COP 21 and the resulting *Paris Agreement* were key events for the sustainability movement and in particular for *green finance*. For the first time, the focus of *Conference of the Parties* was not only on traditional environmental themes, such as the reduction of greenhouse gases emissions and climate change adaptation but also on challenges linked to the financing of the environmental transition. Although funding had always been necessary for any of the global community's initiatives that were linked to the environment, the specific emphasis on the necessity for the participation of financial actors that was given during COP 21, was the first of its kind.

[12] The targets for the first commitment period of the Kyoto Protocol covered emissions of the six main greenhouse gases, namely, carbon dioxide (CO_2), methane (CH_4), nitrous oxide (N_2O), hydrofluorocarbons (HFCs), perfluorocarbons (PFCs) and sulphur hexafluoride (SF_6). The maximum amount of emissions (measured as the equivalent in carbon dioxide) that a Party to the Protocol may emit over a commitment period in order to comply with its emissions target is known as a Party's assigned amount. The individual targets for the Parties are listed in the Kyoto Protocol's Annex B.

[13] The most publicised event on the subject after that was the 2009 conference in Copenhagen (COP 15), when the US President, Barack Obama, a climate action supporter, had just entered office. The event can be considered largely a failure, as results were far below expectations.

[14] Encyclical letter *Laudato sí* of the Holy Father Francis on Care for Your Common Home.

The agreement aimed to strengthen the global response to the threat of climate change by holding the increase in the global average temperature to well below 2 °C above pre-industrial levels and pursuing efforts to limit the temperature increase to 1.5 °C. Additionally, the agreement aimed at increasing the ability of countries to deal with the impacts of climate change, and at making finance flows consistent with a low greenhouse gas emissions and climate-resilient pathways. To reach these ambitious goals, appropriate mobilisation and provision of financial resources, a new technology framework and enhanced capacity-building were also recalled. The agreement also required all Parties to put forward their best efforts through *Nationally Determined Contributions* (NDC) and to strengthen these efforts in the years ahead. This included requirements that all Parties report regularly on their emissions and on their implementation efforts. The Parties also bore a responsibility to meet every five years to discuss the subject and set up a robust, transparent and accountable reporting system to track their progresses. The objectives that were announced during the agreements will be revised in 2020, and once every five years after that initial revision. An overall assessment will be performed in 2023, and, will occur again every five years.

Although the global reach of the negotiations is indisputable, their direct impact on climate change is still somehow contested today. The agreement is only partially legally binding and there are no means to systematically verify if the Parties are reaching their objectives. Some important subjects were also discarded from the debate, including carbon pricing and the possible discontinuation of fossil fuel extractions. Finally, in June 2017, US President Donald Trump announced his intention to withdraw his country from the *Paris Agreement.* Under the agreement itself, the earliest effective date of withdrawal for the United States is November 2020.

1.4 Sustainable Development Goals

In parallel with the preparation of COP 21, the UN was also active in setting up a global framework to encompass every major sustainability-related issue. Much like COP 21 and the *Paris Agreement*, this framework needed substantial preliminary consultations and political efforts before being endorsed by the global community. Earlier in September 2000, world leaders had already gathered on the occasion of the *Millennium Summit* in order to discuss the role of the UN for the next century. It was agreed, inter alia, that the main objective of the organisation was to help members of the world's poorest countries to improve their life conditions. The *Millennium Declaration* was hence drafted, in which the UN General Assembly adopted a series of goals on different sustainability issues. These were synthesised in eight *Millennium Development Goals* (MDG)[15] that were to be reached by 2015. At that time, only one of these goals was directed towards the environment. MDG results were uneven between countries, but the initiative could be considered as a success. Among other results, extreme poverty went from 47% in developing countries in 1990 to 14% in 2015, the number of out-of-school children was halved, child mortality was halved, maternal mortality ratio declined by 45% and new HIV infections fell by 40%. On the subject of ensuring environmental sustainability, 98% of ozone-depleting substances were eliminated since 1990, and millions of people gained access to drinkable water and sanitation.[16] However, the UN also pointed out that important challenges still lay ahead, notably on the subjects of gender and social inequality, conflicts, poverty and hunger, as well as climate change and environmental degradation.

Based on the results attained in 2015, the UN developed new goals for the period 2015–2030. In September 2015, the UN General Assembly and the 193 countries it represented adopted the *2030 Agenda for Sustainable Development*,[17] which was synthesised in 17 goals and 169 targets. These goals were called *Sustainable Development Goals* (SDG). They are listed in Table 1.1. At least nine of these goals can be related to the preservation of the environment or have significant environmental implications. Through the SDG, the UN hence made it clear that protecting the planet is essential to face the needs of present and future generations, and that limiting climate change is necessary to realise this objective as well as protect citizens of the world from environmental catastrophes. With the *Paris Agreement*, a new global framework on sustainable development was hence formed, with a strong focus on the environment.

[15] The MDG were: to eradicate extreme poverty and hunger, to achieve universal primary education, to promote gender equality and empower women, to reduce child mortality, to improve maternal health, to combat HIV/AIDS, malaria and other diseases, to ensure environmental sustainability, to develop a global partnership for development.

[16] For more details, see UN (2015a).

[17] For more details, see UN (2015b).

Table 1.1 Sustainable Development Goals (SDG)

#	*Social Development Goal*	*Short description*	*Environment related*
SDG 1	No poverty	End poverty in all its forms everywhere	
SDG 2	Zero hunger	End hunger, achieve food security and improved nutrition and promote sustainable agriculture	√
SDG 3	Good health and well-being	Ensure healthy lives and promote well-being for all at all ages	
SDG 4	Quality education	Ensure inclusive and equitable quality education and promote lifelong learning opportunities for all	
SDG 5	Gender equality	Achieve gender equality and empower all women and girls	
SDG 6	Clean water and sanitation	Ensure availability and sustainable management of water and sanitation for all	√
SDG 7	Affordable and clean energy	Ensure access to affordable, reliable, sustainable and modern energy for all	√
SDG 8	Decent work and economic growth	Promote sustained, inclusive and sustainable economic growth, full and productive employment and decent work for all	
SDG 9	Industry, innovation and infrastructure	Build resilient infrastructure, promote inclusive and sustainable industrialisation, and foster innovation	√
SDG 10	Reduced inequalities	Reduce income inequality within and among countries	
SDG 11	Sustainable cities and communities	Make cities and human settlements inclusive, safe, resilient and sustainable	√
SDG 12	Responsible consumption and production	Ensure sustainable consumption and production patterns	√
SDG 13	Climate action	Take urgent action to combat climate change and its impacts by regulating emissions and promoting developments in renewable energy	√
SDG 14	Life below water	Conserve and sustainably use the oceans, seas and marine resources for sustainable development	√
SDG 15	Life on land	Protect, restore and promote sustainable use of terrestrial ecosystem, sustainably manage forests, combat desertification and halt and reverse land degradation and halt biodiversity loss	√
SDG 16	Peace, justice and strong institutions	Promote peaceful and inclusive societies for sustainable development, provide access to justice for all and build effective, accountable and inclusive institutions at all levels	
SDG 17	Partnerships for the goals	Strengthen the means of implementation and revitalise the global partnership for sustainable development	

Note: Authors' elaboration based on the SDG description as given in the *2030 Agenda for Sustainable Development* (UN 2015b)

1.5 Green Finance as a (Still) Emerging Asset Class

Today, the framework provided by the *Paris Agreement* and the SDG may also be used to better delineate the perimeter of action of *green finance* (and of other components of the *sustainable finance* landscape). In this respect, *green finance* can be specifically referred to the financial stocks and flows aiming at supporting the achievement of the environment-related SDG. Similarly, *climate finance* can be associated with that component of *green finance* focussing on climate action (in the form of climate change mitigation and climate change adaptation[18]), while *sustainable finance* may be considered to embrace all the financial stocks and flows mobilised to achieve all the SDG. These relations are graphically shown in Fig. 1.1.

Apart from various attempts made to define the term, it can be argued that the appearance of *green finance* as an emerging asset class has been established also due to an unprecedented and continuing increasing engagement of the financial industry along with the political and societal debate surrounding the need for an environmental transition. In fact, following the growing demand for green financial products resulting from the recognition of the risks linked to the degradation of the environment,

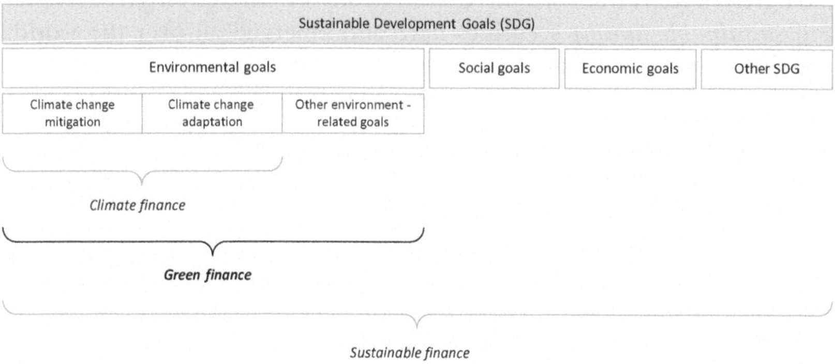

Fig. 1.1 Green finance and its contribution to the Sustainable Development Goals. (Note: Adapted from UNEP 2016)

[18] Climate change mitigation usually refers to efforts to reduce or prevent emission of greenhouse gases (GHG). Climate change adaptation normally concerns the adjustments in ecological, social or economic systems in response to actual or expected climatic modifications and their effects or impacts.

the actors of the financial industry started a structured process of innovation of their offering. The first *green bond* was issued by the European Investment Bank (EIB) in 2007,[19] followed by a similar issuance by the World Bank the year after. Since then, an increasing number of green-labelled products have been created. An "environmental" correspondent for most of the traditional financial products can be found today. In addition, the analysis of specific risks related to climate change gradually became a specific source of operational concern in the financial industry.[20] In this respect, extreme weather-related events due to climate change have forced in particular insurance companies and banks to pay attention to the possible impacts of environmental risks on their bottom line (e.g. in the form of unexpected higher payments on insured risks or higher default rates on loans from companies affected by natural calamities).[21] Today, the growing acceptance of the materiality of climate change at all levels of the economy is another central driver for the development of *green finance*.[22]

Finally, the growth of *green finance* in the last decade should also be related to a strong commitment of the major stock exchanges worldwide. Financial centres such as London, Paris, Luxembourg, Copenhagen, Amsterdam in Europe, Shanghai and Beijing in China, San Francisco and Los Angeles in the United States, Vancouver and Montreal in Canada have taken the lead and are progressively improving the quality and depth of their *green finance* offer. To this extent, dedicated listings for *green finance* and *sustainable finance* securities have also emerged all over the world. Table 1.2 reports the main ones.

[19] This bond was labelled *Climate Awareness Bond.*

[20] Icons like Larry Fink, the chief executive officer of the USD 6.3 trillion asset management firm Blackrock, warned chief executive officers of public companies that "society is demanding that companies, both public and private, serve a social purpose" and expressed his expectation that they start accounting for their impact on society. Robert Litterman, Goldman Sachs' former head of risk, affirmed in an interview that "climate risk is not being priced right by society" and is now a renowned promoter of climate-oriented market solutions.

[21] In this respect, in 2005 hurricane Katrina, which had devastating consequences on the population and had an unprecedented echo in the media, resulted as an event able to trigger major developments for the SRI and ESG industry, as many institutional investors started to invest in ESG research and analytics.

[22] It can be also argued that the development of the SRI and ESG investing industries was linked to the subprime mortgage crisis and the exposure of the great systemic ethical failures of financial markets. Even though the impact of the subprime crisis mostly resulted in social and governance changes rather than environmental, it further raised the question of the extra-financial impact of corporations and financial actors globally.

Table 1.2 Major stock exchanges with dedicated sections for green or sustainable finance

Name of stock exchange	*Type of dedicated section*	*Launch date*
Oslo Stock Exchange	Green bonds	January 2015
Stockholm Stock Exchange	Sustainable bonds	June 2015
London Stock Exchange	Green bonds	July 2015
Mexico Stock Exchange	Green bonds	August 2016
Luxembourg Stock Exchange	Luxembourg Green Exchange	September 2016
Shanghai Stock Exchange	Green bonds	March 2016
Borsa Italiana	Green and social bonds	March 2017
Taipei Stock Exchange	Green bonds	May 2017
Johannesburg Stock Exchange	Green bonds	October 2017
Japan Exchange Group	Green and social bonds	January 2018

Source: CBI (https://www.climatebonds.net/green-bond-segments-stock-exchanges, consulted on 3 February 2019). Luxembourg Green Exchange includes green, social, sustainable and ESG securities

1.6 Main Types of Green Finance Products and Services

1.6.1 Green Bonds (and Climate Bonds)

Green bonds can be considered the most important innovation in the sector of *green finance* since its inception, which paved the way for the development of many other green financial products and services. Since the first issuance by the European Investment Bank in 2007, *green bonds* have been constantly growing in popularity and are now issued by international financial organisations, large corporations, banks and even national governments and municipalities. The issuance of new *green bonds* reached USD 155.5 billion in 2017.[23] *Green bonds* are used by issuers to finance specific projects targeted at creating positive environmental impact.[24] At the same time, buyers of these bonds have the

[23] Source: Climate Bond Initiative (2018).

[24] Based on ICMA (2018), there are currently four types of *green bonds* in the market: *standard green use of proceeds bonds* (standard recourse-to-the-issuer debt obligations aligned with the GBP), *green revenue bonds* (non-recourse-to-the issuer debt obligations aligned with the GBP in which the credit exposure in the bond is to the pledged cash flows of the revenue streams, fees, taxes, etc. and whose use of proceeds go to related or unrelated green projects), *green project bonds* (project bonds for a single or multiple green projects for which the investor has direct exposure to the risk of the projects with or without potential recourse to the

assurance that the proceeds of the issuance are used to these ends. The projects financed by *green bond* issuances typically relate to sectors such as renewable energy, sustainable management of natural resources, energy efficiency (including buildings renewal), pollution control and prevention. As of today, the largest buyers of *green bonds* remain institutional investors.

Specific *Green Bond Principles* (GBP) have been published by the *International Capital Market Association* (ICMA) in 2014 and then updated in 2018,[25] representing the leading framework for defining a *green bond* today. These principles represent voluntary process guidelines that recommend transparency and disclosure and promote integrity in the development of the *green bond* market, by also clarifying the approach for issuance of a *green bond*.[26] To this extent, the main components of use of proceeds, process for project evaluation and selection, management of proceeds, reporting are considered. These principles are also widely followed by green certification agencies, which assure actual and potential investors of the compliancy of each specific issuance to the principles by issuing second–opinions and eventually assigning (or not) green labels.[27]

Climate bonds focus in particular on projects related to climate change mitigation and adaptation. Contrary to *green bonds*, no widespread labelling is currently in use for this category of debt issuances (even though the first labels started appearing in the market). Hence, it can be considered that *climate bonds* today logically include part of the labelled *green bonds* (the ones whose proceeds are used for climate change mitigation and adaptation projects), and a number of unlabelled bonds financing climate action. Contrary to *green bonds*, which are primarily issued by large and well-diversified companies, the *climate bonds* world belongs mostly to pure-play issuers (CBI 2017).

issuer, and that is aligned with the GBP) and *green securitised bonds* (bonds collateralised by one or more specific green projects, including but not limited to covered bonds, asset-backed securities, mortgage-backed securities and other structures and aligned with the GBP and for which the first source of repayment is generally the cash flows of the assets).

[25] See ICMA (2018).

[26] For a more detailed analysis of the GBP, see Chap. 2.

[27] Examples of organisations offering these services are CICERO, Vigeo, Oekom, Sustainalytics, but also Deloitte, EY and KPMG.

1.6.2 Green Asset-Backed Securities

Following the success of the first *green bonds*, more complex structured finance instruments have also started to emerge in the market. These financial products, which are built on the basis of the traditional scheme of securitisation, are usually referred to as *green asset-backed securities.* As asset-backed securities (ABS) are indeed debt instruments, *green asset-backed securities* are often included in the general category of *green bonds.*

Through securitisation, assets that are normally illiquid can be pooled and transferred to an ad hoc vehicle (so-called special purpose vehicle, SPV), which issues tradable securities backed by those assets. Principal and interests for the securities issued by the SPV are then repaid by the flow of resources coming from the underlying pooled assets. The provision of new liquidity, the management of capitalisation through the transfer of risk, the possibility of regulatory arbitrage and the realisation of profits opportunities have been observed in the literature as the key determinants for launching securitisation operations (e.g. Affinito and Tagliaferri 2010; Ahn and Breton 2014; IMF 2015).

ABS can be considered as being "green" when the underlying assets (which normally are not green-labelled) finance environment-related projects, and the issuance of the securities by the SPV is in line with the criteria valid for labelling a *green bond.*[28] Given the diversity in the types of assets that can be securitised and in the structuring methodologies currently in use, the existing *green asset-backed securities* present a high degree of heterogeneity. Securities of this type include mortgage-based structures such as solar ABS, electric automobile ABS, property assessed clean energy[29] ABS, residential and commercial mortgage-backed securities (RMBS and CMBS) and covered bonds. As of today, *green*

[28] Use of proceeds, process for project evaluation and selection, management of proceeds, reporting.

[29] The property assessed clean energy (PACE) model is a mechanism for financing energy efficiency and renewable energy improvements on private property used in the United States. PACE programmes exist for both residential properties (commonly referred to as Residential PACE or R-PACE) and commercial properties (commonly referred to as Commercial PACE or C-PACE).

asset-backed securities make up already a not-negligible proportion of *green bond* issuances, accounting for around 17% of the entire market.[30] Furthermore, it can be expected that the types of underlying assets and the sectors covered will continue to increase in the years to come, further pushing this segment of the market.

1.6.3 Green Loans

As *green bonds*, *climate bonds* and *green asset-backed securities* issuance amounts are necessarily high and the need for funding a wide array of environmental projects requiring smaller-scale investments is indeed not negligible, green lending in the form of *green loans* has also started to be considered. Similar to the GBP, *Green Loan Principles* (GLP) were developed by the *Loan Market Association* (LMA) and the *Asian Pacific Loan Market Association* (APLMA) in 2018.[31] In this respect, the same criteria for use of proceeds, process for project evaluation and selection, management of proceeds, reporting are taken into account for assessing the compliancy of the loans issued with the principles. Labelling services are also available.[32]

Nevertheless, the *green loans* market is still in its early stage of development and it is difficult to predict its possible success at this time. The issuance of a *green loan* (e.g. by a commercial bank to a company) requires substantial screening, labelling, disclosure and control-related activities, which hence represent additional organisational and operational costs. A large part of these costs typically result to be independent from the size of the loan. Henceforth, the economic feasibility of

[30] Source: Moody's, data concerning the first three quarters of 2018.

[31] A green loan is defined as "*any type of loan instrument made available exclusively to finance or refinance, in whole or in part, new and/or existing eligible green projects*". See also LMA (2018).

[32] A similar array of providers of second-opinions acting in the *green bonds* market are also present in the *green loans* sector, in particular CICERO, Vigeo, Oekom and Sustainalytics.

a wide adoption of *green loans* remains highly uncertain at present, and the effective use of *green loans* may remain restricted to loans of big size issued to large companies.[33,34]

1.6.4 Green Funds

The traditional funds industry has also embraced *green finance*. To do that, and without radically changing the way of operating, fund managers have progressively developed a number of approaches to integrate environmental considerations in their investment routines. These approaches, which then characterise a *green fund* (and that are also similar to the ones in use in other branches of *sustainable finance*) refer to[35]:

- Best-in-class strategies, consisting in using specific criteria to identify companies that perform best in terms of environmental impact within different sectors or industries.
- Themed strategies, involving investing in sectors that are essential to climate change and the environment, such as energy efficiency, agriculture or waste management.

[33] Within the broad category of *green loans* several types of products or schemes can indeed be listed. For example, *green commercial building loans* have started to emerge to finance new buildings in line with strict environmental standards or to retrofit existing constructions to improve energy consumption or waste and pollution management. An interesting case study of this type of product is linked to the construction of the *Duo Towers* in Paris. The construction of the two skyscrapers (due for completion in 2020) is funded through the first green-labelled commercial real estate loan in Europe. The EUR 480 million construction is to meet the most rigorous environmental standards, and the buildings' performance in terms of energy consumption and carbon emissions, which will be monitored and reported regularly. Other examples are given by *green mortgages schemes*, which provide clients that wish to purchase energy efficient homes or transform their homes with lower interest rates than plain loans or by the *Energy Efficient Mortgage Action Plan* (EeMAP) initiative launched in June 2018 by the *World Green Building Council* in partnership with major European banks such as BNP Paribas, Société Générale, Nordea Bank and ING Bank. Smaller-scale financial products can in principle also be designed, in particular for smaller, more widespread funding needs. *Green car loans* are already conceptualised and may offer preferential rates for less polluting vehicles.

[34] In case of *green loans* dedicated to large companies, at least three technical possibilities may be foreseen: green bilateral loans (a contract formalised by a bank and a company), syndicated loans (in which a group of banks finance a company for a specific project) or a green revolving credit line (in which a bank gives to a company the availability of funds for future projects and activities which are in line with the GLP but are not defined *ex-ante*).

[35] See also Eurosif (2018) and Swiss Sustainable Finance (2017).

- Norms-based strategies, relying on national or international standards and norms to build green portfolios.
- Engagement and voting strategies, implying investing in companies and actively participating in company's decisions-making encouraging better environmental management and processes.
- Exclusion strategies, consisting in excluding companies from portfolios that do not suit given environment-related principles.

These funds, which can invest both in equity and debt instruments, hence provide their clients with platforms through which environmentally friendly businesses and organisations are supported with long-term funding.[36] Their growing popularity amongst retail investors in particular is given by the ease of buying and selling quotas of the funds.[37]

1.6.5 Green Project Financing Operations

Project finance generally refers to different techniques of financing a project for which the repayment is directly dependent on future cash flows generated by the initiative (and not linked to the creditworthiness of its promoters). Project financing operations are extensively used when it comes to large, long-term infrastructure and industrial projects or other types of non-recurring initiatives. These operations usually involve a number of equity investors (the sponsors) and a pool of banks or other financing institutions (the syndicate). In this respect, several types of arrangements and contractual relationships exist (e.g. Yescombe 2002).

Following the overall development of the *green finance* market, dedicated green project finance teams have been created by commercial and investment banks throughout the world. In principle, what eventually characterises a project financing operation as "green" is the use of pro-

[36] In recent years, several multinational development banks, including the World Bank and EIB announced their commitment to increasing capital flows directed to support the transition to a low-carbon economy. In doing that, they can also establish specific funds. For instance, the IFC (*International Financial Corporation*, member of the World Bank Group) initiated in 2017 the work on Amundi's new USD 2bn *Cornerstone Green Bond Fund*, which will invest in bonds issued by banks in emerging markets. The IFC has committed up to USD 325 ml and the EBRD intends to invest up to USD 100 ml.

[37] Labelling services for investment fund are also emerging. An example of providers of these services is Luxflag.

ceeds coming from the financing deal between the sponsors and the syndicate. As of today, in most of the cases *green project financing operations* have been organised for large-scale projects in the renewable energy sector. Nevertheless, a standard definition and a target structure for *green project financing operations* have not been elaborated so far from the financial industry. In practice, the identification of one operation as "green" may result from an autonomous decision of the sponsors. Only a few principles have recently emerged,[38] but a widespread acceptance has not been observed yet.

1.6.6 Green Indices

Market indices play an important role in the modern financial markets. As they provide cost-effective, easy-to-understand and usable information on the products they are built on, they effectively contribute to drive the demand and supply of securities among both institutional and retail investors. In this respect, indices are also used by the funds industry as benchmarks for investment strategies.

Green indices have recently flourished (particularly from 2014) and are today present in all the most important financial centres. They can focus on specific types of securities (in particular on fixed income and equity), on specific sectors (such as water management, solar energy or renewable energies), or they may be built to represent the wider green sector. To date, a great majority of traditional indices providers have developed sustainability-oriented or green equivalents.[39]

[38] The *Equator Principles* are an example. Formulated by the *Equator Principles Association* (EPA), they represent a risk management framework, aimed at financial institutions, for determining, assessing and managing environmental and social risks in projects. Such a framework is primarily intended to provide a minimum standard for due diligence and monitoring to support risk decision-making. The *Equator Principles* apply to project finance (including advisory services), project-related corporate loans and bridge loans.

[39] A few examples of green indices are the *S&P Green Bond Index* (focused on green bonds), the *S&P Green Project Bond Index* (designed to capture bonds which produce environmental benefits but don't necessarily carry green labels) or the *S&P 500 Environmental & Socially Responsible Index* (which measures the performance of securities from the S&P 500 that meet specific social and environmental criteria).

1.7 Relevant Market Trends in Green Finance

Green bonds have historically contributed to the development of the *green finance* market. Nevertheless, a sustained increase in the level of global issuance can be observed only starting from 2014 (see Fig. 1.2), when it can be stated that the market of green securities concluded its phase of inception and entered in a phase of first growth (which is still ongoing in 2019). In terms of geographical impact, the United States, China and the largest European countries have so far led the way in terms of new and overall issuance (see Fig. 1.3).

With the growth of the *green bonds* market, a wide spectrum of sectors financed can also be observed. Even though the largest share of projects financed refers to "traditional" green sectors such as renewable energy, low-carbon buildings, energy efficiency and clean transport, an increasing diversification is progressively arising, with proceeds from *green bonds* being used at present in sustainable waste management, sustainable land use and forestry and climate change adaptation (see Fig. 1.4).

The funds industry has also progressively consolidated its presence in the *green finance* market, in particular, thanks to the expansion of an increasing availability of green-labelled and environmentally aligned

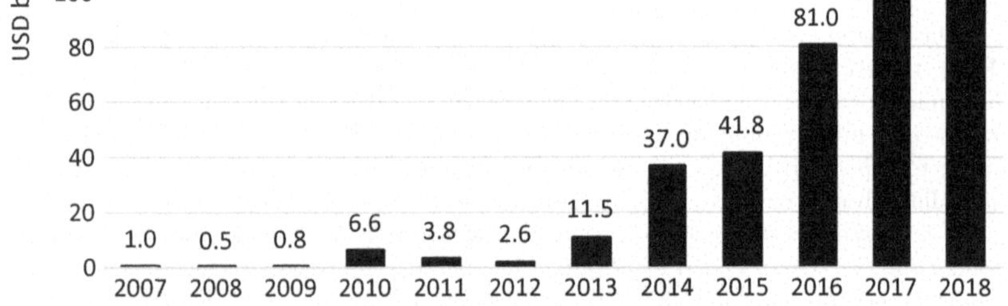

Fig. 1.2 Issuance of green bonds by year. (Note: Authors' elaboration on data Bloomberg, CBI and European Investment Bank. Values refer to labelled securities and include green bonds and ABS. Data for 2018 is an estimation)

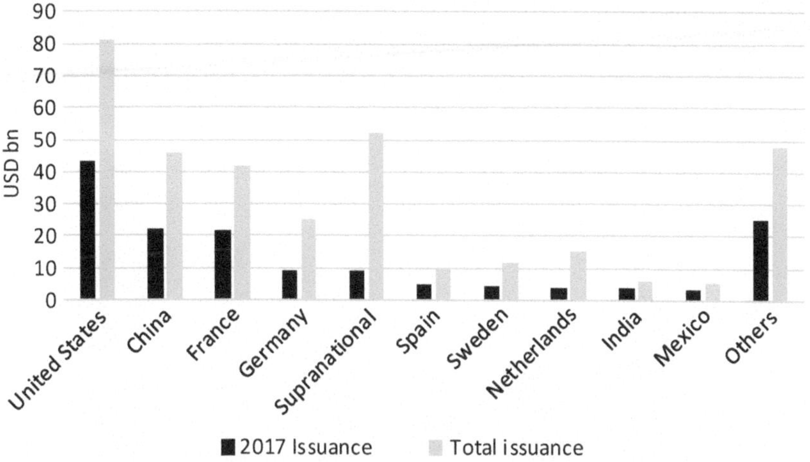

Fig. 1.3 Green bonds issuance by country. (Source: Authors' elaboration on data CBI. Values refer to labelled securities and include green bonds and ABS)

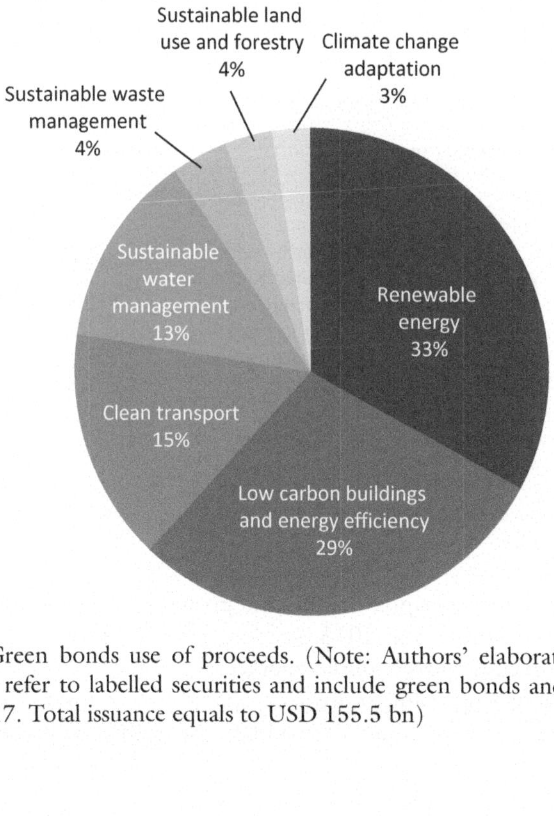

Fig. 1.4 Green bonds use of proceeds. (Note: Authors' elaboration on data CBI. Values refer to labelled securities and include green bonds and ABS. Data refers to 2017. Total issuance equals to USD 155.5 bn)

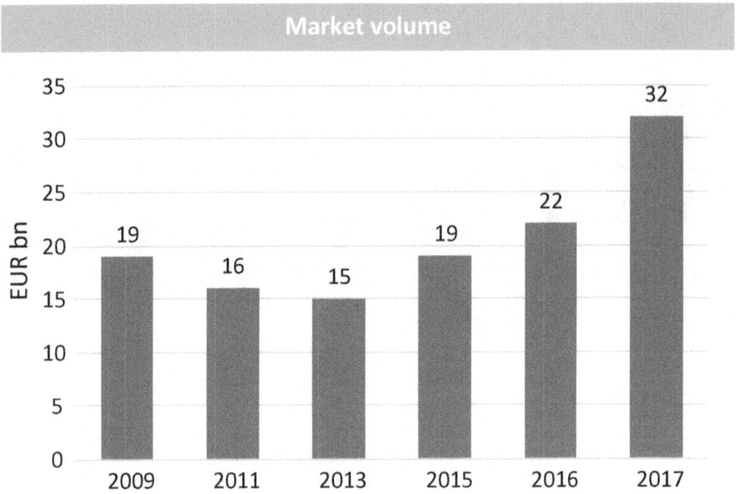

Fig. 1.5 Green funds in Europe: market volumes. (Source: Adapted from Novethic 2018)

Table 1.3 Green funds in Europe: main players

Top 10	*Fund management company*	*Country*	*Number of funds*	*AUM (€ bn)*
1	Pictet AM	Switzerland	3	5867
2	BNP Paribas AM	France	10	5232
3	Amundi	France	8	2349
4	Blackrock	United Kingdom	5	1845
5	RobecoSAM	Switzerland	2	1392
6	Triodis IM	Netherlands	3	1114
7	ÖkoWorld	Germany	5	1016
8	ASN Bank	Netherlands	2	951
9	KBC AM	Belgium	6	842
10	Swisscanto	Switzerland	7	784
Total			**51**	**21,392**

Source: Adapted from Novethic (2018). Data on AUM (assets under management) refers to 2017

securities. In Europe, signs of relevant growth for green funds can be observed in particular from 2013 (see Fig. 1.5) (Table 1.3).

The growth of the *green finance* market can hence be considered robust today, and increasing volumes are accompanied by sectorial diversification and sustained by a wide range of products. Nevertheless, it can be argued that the actual levels of emissions of green securities are still nothing more than "a drop in the ocean" when they are compared to estimated needs for an effective financing of the environmental goals.[40]

1.8 Open Issues and Main Challenges Ahead[41]

Despite the positive momentum that has been featuring the *green finance* market for the past years, some significant challenges ahead can be listed. Yet, the way these challenges will be faced, by both industry and policy makers, will have a decisive impact on the development potential of *green finance* in the years to come.

1.8.1 Better Defining "Green" in Finance

A common understanding of what is a "green" activity or investment is indeed of fundamental importance to attract capital to these investments. As of today, several definitions of "green" are in use in the financial market. The existing definitions have mainly been created by practitioners to segregate green securities from their non-green equivalents. They are mainly backed by criteria identifying in particular the possible use of the proceeds coming from the specific financing instrument and other operational standards (as in the case of *green bonds* and *green loans*), or the eligible activities carried out by the companies financed (as in the case of *green funds*). Nevertheless, high heterogeneity can be observed as concerns approaches, scopes and reliability of the organisations issuing these definitions. In the longer term, this lack of clarity on *what is green* and *what is not green* may harm the credibility of the market. In this respect, an effort should be encouraged at international level to produce reliable and widely accepted references. In Europe, a structured initiative in this direction has been launched by the European Commission.[42]

[40] As an example, investments of around EUR 520–575 bn annually have been estimated to be necessary in the EU only in order to achieve a net-zero greenhouse gas economy in the 2050 horizon. Source: EC (2018b).

[41] This paragraph only summarises the main open issues and challenges for *green finance*. A wider dissertation is given all along the rest of book, and in particular in Chaps. 2, 5, 6, 7 and 9.

[42] In May 2018, the European Commission adopted a proposal for a regulation on the establishment of a framework to facilitate sustainable investment. This regulation establishes the conditions and the framework to gradually create a unified classification system ("taxon-

1.8.2 Better Understanding Market Incentives to Issue Green Securities

Financial incentives for market players to issue green securities are today still difficult to assess. Even though at least reputational gains and corporate social responsibility acknowledgement by existing and prospective clients can be mentioned as possible paybacks for embracing *green finance*,[43] a lot of research is still needed to evaluate whether direct financial benefits exist to issue green securities. Examples of these benefits may be a lower cost of green-labelled debt (vis-à-vis a corresponding unlabelled one), faster placement of green securities in the portfolios of institutional investors or a reduction in the impact of climate-related risks on the business of issuing organisations. This question is of high relevance as the possible lack of concrete financial incentives for potential issuers of green securities may represent in the longer term a substantial hurdle to the further development of the market. In this respect, market research needs to be encouraged in order to bring useful evidence to investors, issuers and policy makers.

1.8.3 Mainstreaming Green Finance for a Real Impact on the Environment

The inherent policy objective of *green finance* is to contribute to foster an environmentally sustainable economy. In this respect, it can be stated that for *green finance* to effectively contribute to face climate change and to fully support other environmental objectives, a shift in paradigm should occur. From a promising niche, *green finance* should evolve into a mainstream way of financing the economy. To do that, only the introduction of a structured policy framework can be effective. This framework should be in particular focused on fostering a strong demand and supply of green securities and encouraging the full inclusion of environmental risks in the

omy") on what can be considered an environmentally sustainable economic activity. Following this proposal for a regulation, in June 2019 the Technical Expert Group (TEG) established by the European Commission to support (inter alia) in the definition of the taxonomy, released its technical report (see TEG 2019). This report will be a key element in the establishment of the first version of the taxonomy. For a more detailed dissertation, see Chap. 6.

[43] Abundant evidence exists showing a generally positive relationship at least in the last 20 years between corporate finance performances and environmental, social and governance (ESG) performance. See for example Friede et al. (2015).

investors' decision-making processes. As of today, preliminary concrete actions in this direction have been taken in particular in the EU, with the *Action Plan for a greener and cleaner economy*[44] and follow-up initiatives.[45] Nevertheless, in the best case, several years will be needed before the first results will eventually occur.

1.8.4 Facing Intermittent Political Commitment

Political commitment towards environmental sustainability has been historically intermittent. Nevertheless, it is increasingly evident that the fortune of *green finance* will decisively depend on the level of engagement of the international community towards the environmental goals (particularly in the framework of the *Paris Agreement* and the SDG). In this respect, very different appreciations can be observed, in particular between the two sides of the Atlantic. On the one hand, the EU has taken the lead and has already tabled concrete initiatives to foster the change. The issuance of the *European strategic long-term vision for a prosperous, modern, competitive and climate neutral economy*[46] is a concrete example of this engagement. On the other hand, announcement of the retirement of the United States from the *Paris Agreement* signals an inversion of course and may significantly weaken the international action for the achievement of the environmental goals. As a matter of fact, political disengagement with respect to the environment may harm the possibilities of development of *green finance* already in the near future.

[44] See EC (2018a).

[45] In particular, following the *Action Plan for a greener and cleaner economy*, in May 2018 the European Commission adopted a package of implementing measures including: a proposal for a regulation on the establishment of a framework to facilitate sustainable investment (this regulation establishes the conditions and the framework to gradually create a unified classification system—or taxonomy—on what can be considered an environmentally sustainable economic activity); a proposal for a regulation on disclosures relating to sustainable investments and sustainability risks and amending Directive (EU)2016/2341 (this regulation aims at introducing disclosure obligations on how institutional investors and asset managers integrate environmental, social and governance factors in their risk processes); a proposal for a regulation amending the benchmark regulation (to create a new category of benchmarks comprising low-carbon and positive carbon impact benchmarks, which should provide investors with better information on the carbon footprint of their investments). See also Chap. 6.

[46] EC (2018b).

References

Affinito, M., & Tagliaferri, E. (2010). Why do (or did?) banks securitize their loans? Evidence from Italy. *Journal of Financial Stability, 6*, 189–202.

Ahn, J. H., & Breton, R. (2014). Securitization, competition and monitoring. *Journal of Banking and Finance, 40*, 195–210.

Ayadi, R., Llewellyn, D., Schmidt, R. H., Arbak, E., & De Groen, P. W. (2010). *Investigating diversity in the banking sector in Europe: Key developments, Performance and Role of Cooperative Banks*. Brussels: CEPS Paperbacks.

Ayub, M. (2013). *Understanding Islamic finance*. Hoboken: Wiley.

Climate Bonds Initiative (CBI). (2017, September). *Bonds and climate change. The state of the market*. London.

Climate Bonds Initiative (CBI). (2018, January). *Green Bonds Highlights 2017*. London.

European Commission (EC). (2018a). *Action plan: Financing sustainable growth*, COM(2018) 97 final. Brussels.

European Commission (EC). (2018b). *A clean planet for all. A European strategic long-term vision for a prosperous, modern, competitive and climate neutral economy*, COM(2018) 773 final. Brussels.

European Sustainable Investment Forum (Eurosif). (2018). *European SRI study 2018*. Brussels.

Friede, G., Busch, T., & Bassen, A. (2015). ESG and financial performance: Aggregated evidence from more than 2000 empirical studies. *Journal of Sustainable Finance and Investment, 5*(4), 210–233.

Glaeser, L., & Scheinkman, J. (1998). Neither a borrower nor a lender be: An economic analysis of interest restrictions and usury laws. *Journal of Law and Economics, 41*(1), 1–36.

Intergovernmental Panel on Climate Change (IPCC). (2018). *Special Report. Global warning of 1.5 °C*. Geneva.

International Capital Market Association (ICMA). (2018, June). *Green bond principles. Voluntary process guidelines for issuing Green Bonds*. Zurich.

International Monetary Fund (IMF). (2015, January). *Securitization: The road ahead*, IMF Staff discussion note. Washington, DC.

Lewison, M. (1999). Conflicts of interests? The ethics of usury. *Journal of Business Ethics, 22*(4), 327–339.

Loan Market Association (LMA). (2018). *Green loan principles*. London.

Markowitz, H. (1952). Portfolio selection. *The Journal of Finance, 7*(1), 77–91.

Migliorelli, M. (Ed.). (2018). *New cooperative banking in Europe. Strategies for adapting the business model post-crisis*. London: Palgrave Macmillan.

Novethic. (2018, April). *The European green funds market 2018*. Paris.

Renneboog, L., ter Horst, J., & Zhang, C. (2008). Socially responsible investments: Institutional aspects, performance, and investor behaviour. *Journal of Banking and Finance, 32*(9), 1723–1742.

Swiss Sustainable Finance. (2017, December). *Handbook on sustainable investments. Background information and practical examples for institutional asset owners.* Zurich.

Technical Expert Group on Sustainable Finance (TEG). (2019, June). *Taxonomy.* Technical Report. Brussels.

United Nations (UN). (2015a, July). *The millennium development goals report.* New York.

United Nations (UN). (2015b). *Transforming our world: The 2030 agenda for sustainable development.* New York.

United Nations Environment Programme (UNEP). (2005, October). *A legal framework for the integration of environmental, social and governance issues into institutional investment.* Geneva.

United Nations Environment Programme (UNEP). (2016, September). *Definitions and concepts. Background note.* Inquiry working paper 16/13. Geneva.

Warde, I. (2000). *Islamic finance in the global economy.* Edinburgh: Edinburgh University Press.

Yescombe E. R. (2002). *Principles of project finance.* San Diego, CA: Elsevier.

CHAPTER 2

Defining Green Finance: Existing Standards and Main Challenges

Romain Berrou, Nicola Ciampoli, and Vladimiro Marini

2.1 Introduction

As of today, a unique, universal, commonly accepted definition of *green finance* does not exist. Neither does it for the different families of green financial securities, products or services. In the best cases, industry guidelines have emerged and consolidated through time only for some key categories of securities (e.g. for *green bonds*) and are considered as effective but non-binding market references.

In general terms, the issue of defining *green finance* needs understanding of two separate aspects. The first concerns the sectors or activities that can be financed with green funds. The second regards the operational standards that need to be followed for labelling a specific security, product

R. Berrou (✉)
IAE Université Paris 1 Panthéon-Sorbonne (Sorbonne Business School), Paris, France
e-mail: Romain.Berrou@etu.univ-paris1.fr

N. Ciampoli
LUMSA University, Rome, Italy

V. Marini
University of Rome Tor Vergata, Rome, Italy
e-mail: vladimiro.marini@uniroma2.it

M. Migliorelli, P. Dessertine (eds.), *The Rise of Green Finance in Europe*, Palgrave Studies in Impact Finance,
https://doi.org/10.1007/978-3-030-22510-0_2

or service as green. The first aspect implies a thorough analysis of the impacts of each sector or activity on the environment, including but not limited to their contribution to *climate change mitigation* and *climate change adaptation*.[1] In this respect, if some sectors and activities can be easily considered as being fully eligible for green financing (e.g. in the case of renewable energy or energy efficiency projects), for others this assessment may not be straightforward (e.g. in the case of hybrid vehicles or diesel railways). In the current industry practices different taxonomies today exist listing each time the sectors or the activities considered to be entitled to green financing. As concerns the second aspect, that is specific operational standards to be followed for considering a security, product or service as green, several methodologies have been progressively proposed by the industry in order to attract investors' appetite for specific products categories (e.g. in the case of *green funds*, *green bonds*, *green loans* or even *green indices*). In particular in the case of securities, these standards generally refer to the processes for evaluation and selection of the projects to be financed, the way proceeds are managed and the characteristics of the reporting. These elements are hence assessed by specialised service providers (vis-à-vis ideal benchmarks) in order to decide the eligibility for a green label.

In point of fact, labelled green financial securities, products or services today represent the core components of the *green finance* market. Nevertheless, a wider interpretation of the market is often given, and *green finance* can be considered to encompass also unlabelled financial securities, products or services, when the corresponding financial flows and stocks indeed contribute to finance environmental sustainability (see also Fig. 2.1).

The issue of clearly defining *green finance* is not a secondary one. On the contrary, it is central to the debate surrounding the future of the market. Three main reasons should be highlighted to explain the critical role played by any possible definition of *green finance*. First, identifying what is green implies a decision on which are the sectors or the activities that are

[1] *Climate change mitigation* usually refers to efforts to reduce or prevent emission of greenhouse gases (GHG). *Climate change adaptation* normally concerns the adjustments in ecological, social or economic systems in response to actual or expected climatic modifications and their effects or impacts.

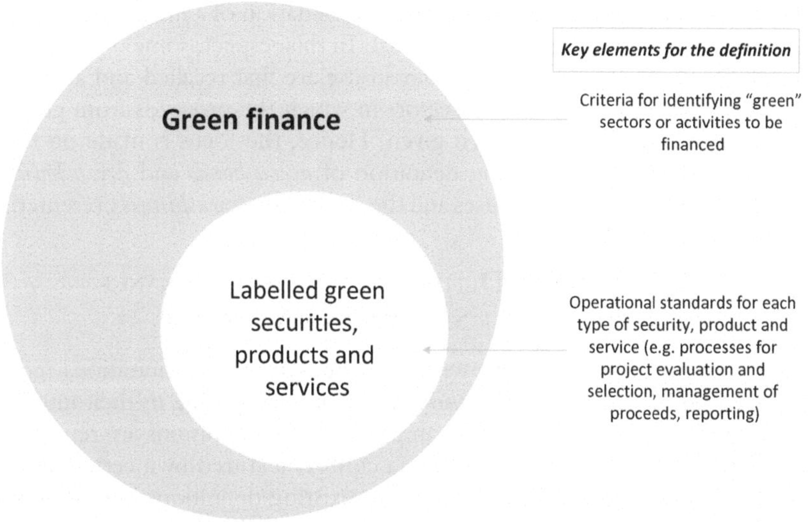

Fig. 2.1 Defining green finance, general framework. (Note: Authors' elaboration)

considered to contribute to reach the environmental goals. In this respect, only green sectors or activities will benefit from the (probably growing) demand-driven flow of resources mobilised by *green finance*, while all the others may remain excluded. Second, on the basis of a specific definition of what is green in the financial market, specific policy measures could eventually be taken to encourage the environmentally friendly investment, such as lower capital requirements for financial intermediaries that hold green securities or ad hoc fiscal incentives. This would again result in specific benefits for some sectors or activities, in particular in terms of higher amount of resources available or lower cost of financing. Third, uncertainty on what green means in the financial sector can significantly harm the credibility of the nascent market and represent in the mid-term, a substantial hurdle to its development. In fact, environmentally conscious economic agents would accept to enter the market only if there is assurance that their investment decisions align to the use of their money for which it is meant.

In such a framework, the aim of this chapter is to provide an overview of the essential elements featuring the definition of *green finance* and to discuss the hurdles still hampering the consolidation of a unique, universal and commonly accepted set of standards. In this respect, some of the leading definitions of *green finance* today in use are first recalled and a structured classification of the main sectors in which the proceeds from green instruments are invested is then given. Hence, the focus is made on the current industry-based leading definition of *green bonds* and *green loans*. Finally, a discussion on the causes and the risks of *greenwashing* is presented.

2.2 Possible Definitions of Green Finance and Eligible Sectors and Activities

Financial institutions, governments and international organisations tend to define *green finance* (and *sustainable finance*) according to their underlying motivations (IFC 2017). Examples of these definitions are reported in Table 2.1. In this respect, even in a context featured by a certain level of heterogeneity, it can be observed that existing definitions indeed present significant similarities, in particular as concerns the characteristic inclusion of references to the protection of the environment and to the financing of the environmental transition.[2,3] The major elements of differentiation in the existing approaches to *green finance* are on the contrary usually entrenched in the operational routines adopted by the different financing bodies and refer first and foremost to the possible interpretation of the impact of specific sectors or activities on the environment. Largely independent from the adoption of a formal definition of *green finance*, many financial institutions, governments and international organisations have to this extent established processes to ensure that environmental criteria are considered for each project to which they allocate green funds. In many cases, this has implied the development of specific taxonomies listing the sectors or activities eligible for these funds.[4] Table 2.2 summarises the

[2] Similar elements are included in definitions developed by non-international organisation or governments. For example, Höhne et al. (2012) define *green finance* as "a broad term that can refer to financial investments flowing into sustainable development projects and initiatives, environmental products, and policies that encourage the development of a more sustainable economy. *Green finance* includes *climate finance* but is not limited to it. It also refers to a wider range of other environmental objectives, for example industrial pollution control, water sanitation, or biodiversity protection".

[3] In addition, it is worth mentioning that no reference is usually made to labelling requirements in the existing definitions of *green finance*.

[4] Among the most important taxonomies (or similar catalogues) should be mentioned the Bank of China's *Green Bond Endorsed Project Catalogue*, the eligibility criteria of EIB's

Table 2.1 Examples of existing definitions of green finance (and sustainable finance)

Organisation	*Definition of green finance or sustainable finance*
Organisation for Economic Co-operation and Development (OECD)	*Green finance* is finance for achieving economic growth while reducing pollution and greenhouse gas emissions, minimising waste and improving efficiency in the use of natural resources.
Government of Germany	*Green finance* is a strategic approach to incorporate the financial sector in the transformation process towards low-carbon and resource-efficient economies, and in the context of adaptation to climate change.
People's Bank of China (PBoC)	*Green finance* policy refers to a series of policy and institutional arrangements to attract private capital investments into green industries such as environmental protection, energy conservation and clean energy through financial services including lending, private equity funds, bonds, shares and insurance.
European Commission (EC)	*Sustainable finance* generally refers to the process of taking due account of environmental and social considerations in investment decision-making, leading to increased investments in longer-term and sustainable activities. More specifically, environmental considerations refer to climate change mitigation and adaptation, as well as the environment more broadly and related risks (e.g. natural disasters). Social considerations may refer to issues of inequality, inclusiveness, labour relations, investment in human capital and communities.
Swiss Federal Ministry of Environment (FOEN)	*Sustainable finance* is defined as financial products and services, under the consideration of environmental, social and governance factors throughout the whole risk management and decision-making process, provided to promote responsible investments which create a positive environmental, social and governance impact.
Indonesian Financial Services Authority (OJK)	*Sustainable finance* is defined as a comprehensive support from the financial service industry to achieve sustainable development resulting from a harmonious relationship between economic, social and environmental interests.

Source: Adapted from UNEP (2016)

Table 2.2 Economic sectors or activities and treatment in existing green finance taxonomies

Main category	*Treatment of the sector or activity in existing green finance taxonomies*		
	Core element	Additional element	Questioned element
Clean energy	Solar energy Wind energy (in-shore and off-shore) Geothermal energy Small hydropower Biomass energy	Waste-to-energy systems Cogeneration Other renewables	Nuclear energy Large hydropower Bioenergy feedstock production Clean coal Improvements in fossil fuels Cleaner fuel production
Energy transmission and storage	Transmission systems for renewables Storage systems Smart grids and mini grids	Improving efficiency of transmission systems	*None*
Energy efficiency	Waste heat recovery Industrial energy efficiency systems Energy efficiency products	Energy efficient products	Energy efficiency in fossil fuel use
Transport	Urban mass transit Non-diesel railways	Electric and hybrid vehicles Alternative fuel vehicles Bicycle, pedestrian, waterways logistics improvement	Diesel railways Rail for transport of fossil fuels
Green buildings	Building retrofitting New green buildings Energy audits and energy services Equipment improvement (e.g. lights, heating, ventilation, air conditioning)	Advanced materials	*None*
Agriculture and land use	Energy and water saving Afforestation and plantations Reforestation Sustainable forest management	Conservation agriculture Sustainable fisheries Identification of protected ecosystems Ecotourism	*None*

Water management	Water saving	Municipal water management Industrial and agricultural water supply Improved drainage Treatment of wastewater to meet compliance obligations	*None*
Non-energy greenhouse gases (GHG) emissions	Coal mine methane capture Carbon capture and storage (CCS) Reduction in GHGs (e.g. in cement, chemicals)	*None*	*None*
Pollution control and waste management	Air/water pollution control Soil remediation and mine rehabilitation Waste gasification Composting Recycling	*None*	Landfill and incineration of waste without energy or gas capture
Disaster prevention and economic resilience	*None*	Climate resilient infrastructures Early warning systems Insurance against natural disasters	*None*
Other sectors	*None*	Broadband Data centres using renewable energy Low-carbon energy-powered mobile base stations Virtual conferencing and tech substitution	*None*

Source: Adapted from UNEP (2016). *Core element* means that the sector is commonly included in the main existing green finance taxonomies. ***Additional element*** means that the sector is included in some of the existing taxonomies. ***Questioned element*** means that more work may be needed to determine specific standards for the inclusion of the sector in any taxonomy. The table does not aim to be exhaustive

treatment typically given to the main economic sectors and activities in the existing *green finance* taxonomies.

Some sectors and activities are universally considered as being fully eligible for green funding, as for example in the case of the main sources of renewable energy (e.g. solar or in-shore and off-shore wind), energy storage systems, smart grids or air and water pollution control systems. On the contrary, some sectors and activities are included only in some of the existing taxonomies and not in others, on the basis of a stricter interpretation of their positive environmental impact. This is the case for example for waste-to-energy systems, electric and hybrid vehicles, eco-tourism or climate resilient infrastructures. Finally, more work and analysis appear to be still needed to determine specific criteria for assessing the eligibility for green funding for some other sectors and activities, for which the available environmental impact analyses may remain somehow incomplete (in particular when primarily based on the assessment of contribution of the sector or activity to the reduction of the greenhouse gas emission, without considering possible side effects). This is the case for example of nuclear energy, diesel railways or bioenergy feedstock production.

The need for common eligibility criteria for green funding between the different financial institutions, governments and international organisations is progressively materialising. Considering the possibility of easily exchanging securities in the financial markets worldwide, different interpretations of what is green may affect the confidence of investors and hence the growth potential of the market. As a matter of fact, a global governance on this issue would be needed in order to coordinate the

Climate Awareness Bonds, the *Common Principles for Climate Mitigation Finance Tracking* developed by a group of multilateral development banks (MDB) together with the International Development Finance Club (IDFC), the sectors eligible for use of proceeds in the framework of *Green Bond Principles* as developed by the *International Capital Market Association* (ICMA) and the *Climate Bond Taxonomy* developed by the *Climate Bond Initiative*. See also MDB (2015), CBI (2018a), ICMA (2018) and, for the Bank of China's *Green Bond Endorsed Project Catalogue*: http://www.greenfinance.org.cn/displaynews.php?cid=79&id=468

efforts of different jurisdictions and private organisations currently active in the development of these criteria.[5,6]

[5] In this respect, a particularly relevant comparison of the standards used in *green finance* taxonomies has been recently developed conjointly by the European Investment Bank (EIB) and the Green Finance Committee (GFC) of the China Society for Finance and Banking in the view of creating the conditions for a possible progressive harmonisation (see EIB and GFC 2017). These institutions compared the different use-of-proceeds classifications under the Bank of China's *Green Bond Endorsed Project Catalogue*, the eligibility criteria of EIB's *Climate Awareness Bonds* and the *Common Principles for Climate Mitigation Finance Tracking* developed by a group of multilateral development banks (MDB) together with the International Development Finance Club (IDFC). For the latter, the multilateral development banks involved were: the African Development Bank (AfDB), the Asian Development Bank (ADB), the European Bank for Reconstruction and Development (EBRD), the EIB, the Inter-American Development Bank Group (IDBG) and the World Bank Group (WBG). The MDB and the IDFC are hereafter referred to as MDB-IDFC. The technical conclusions of the EIB and GFC study were:

- The Chinese, MDB-IDFC and EIB standards use different categories for the classification of the underlying assets. While the Chinese *green bond* catalogue, which is largely consistent with the *Green Bond Principles* (see Sect. 2.3 in the chapter), has a broader scope of green, covering "environmental protection" among others, the MDB-IDFC and EIB standards are focused on "climate change". However, both standards include areas not included in the other.
- Regarding the Chinese standard, within "climate change mitigation", four categories are not included in the MDB-IDFC standard, namely energy saving on greenfield facility construction for industries with national energy consumption allowance, clean utilisation of coal, ultra-high voltage grid infrastructure as well as urban underground pipeline projects. On the other hand, within the broader scope of the Chinese standard, some items outside the MDB-IDFC standard are included, namely environmental restoration projects, coal washing and processing for the purpose of clean utilisation of coal, cleaner gasoline and diesel and a few aspects of ecological protection and climate change adaptation. These differences are similar between the Chinese and the EIB standard.
- When it comes to the EIB standard, as "climate change mitigation", i.e. "low carbon", is the scope of both the MDB-IDFC and the EIB standard, the difference between the two lies in what specific categories to cover within such scope. Here the analysis finds that the EIB lending standard is different from the MDB-IDFC standard in its inclusion of nuclear energy. This difference also exists in the Chinese standard, which does not include nuclear energy either.
- The MDB-IDFC standard further includes a number of categories not included in the Chinese or EIB standard. As opposed to the Chinese standard, the MDB-IDFC standard specifically includes renewable energy power plant retrofits, wind-driven pumping systems, energy audits to end-users, carbon capture and storage, non-motorised transport, projects producing low-carbon components, as well as a number of aspects of technical assistance. Lastly, the MDB-IDFC standard also includes categories not included in the EIB standard, namely energy efficiency in thermal power stations (coal). Energy efficiency in conventional coal-fired power plants is ineligible for EIB unless it meets specific emission performance standards and is in all cases not counted as "climate mitigation".

[6] At the EU level, an action has been recently launched by the European Commission. European Commission aiming at establishing an EU classification system (or taxonomy) for

2.3 Definition of Green Bonds

When it comes to defining specific securities, the framework put in place for *green bonds* is by far the most advanced one. This framework, developed within the financial industry, today benefits from a large acceptance of the *Green Bond Principles* (GBP), issued by the *International Capital Market Association* (ICMA) in 2014 and then updated in 2018. The GBP are voluntary process guidelines that recommend transparency and disclosure and promote integrity in the development of the *green bond* market by clarifying the approach to be followed for the issuance of a *green bond* (ICMA 2018). To this extent, *green bonds* are first defined, as "any type of bond instrument where the proceeds will be exclusively applied to finance or refinance, in part or in full, new and/or existing eligible *green projects* and which are aligned with the four core components of the GBP".[7] The GBP then provide issuers with guidance on the four core components involved in launching a *green bond*, that is use of proceeds,[8] process for project evaluation and selection, management of proceeds and reporting. Table 2.3 details these guidelines.

The framework provided by the GBP hence recommend a structured process for issuers, which investors, banks, underwriters, placement agents and others may use to appreciate the expected features of any given *green bond* (ICMA 2018). The GBP emphasise in particular the required transparency, accuracy and integrity of information that will be disclosed and

sustainable activities is a step in this direction. This initiative resulted in a Report of the Commission Technical Expert Group (TEG) providing a proposed taxonomy for "climate mitigation" and "climate change adaptation" activities (see TEG 2019). For a more detailed discussion, see in particular Chap. 6.

[7] ICMA (2018) also proposes a classification of the different types of *green bonds* in the market: *standard green use of proceeds bonds* (standard recourse-to-the-issuer debt obligations aligned with the GBP), *green revenue bonds* (non-recourse-to-the issuer debt obligations aligned with the GBP in which the credit exposure in the bond is to the pledged cash flows of the revenue streams, fees, taxes etc. and whose use of proceeds go to related or unrelated green projects), *green project bonds* (project bonds for a single or multiple *green projects* for which the investor has direct exposure to the risk of the projects with or without potential recourse to the issuer, and that is aligned with the GBP) and *green securitised bonds* (bonds collateralised by one or more specific *green projects*, including but not limited to covered bonds, asset-backed securities, mortgage-backed securities and other structures and aligned with the GBP and for which the first source of repayment is generally the cash flows of the assets).

[8] As a matter of fact, the GBP suggest an additional taxonomy (even though not intended to be complete or mandatory).

Table 2.3 Core components of the green bond principles

Component	*Guidelines*
Use of proceeds	The cornerstone of a ***green bond*** is the utilisation of the proceeds of the bond for ***green projects***, which should be appropriately described in the legal documentation for the security. All designated ***green projects*** should provide clear environmental benefits, which will be assessed and, where feasible, quantified by the issuer. In the event that all or a proportion of the proceeds are or may be used for refinancing, it is recommended that issuers provide an estimate of the share of financing vs. refinancing, and where appropriate, also clarify which investments or project portfolios may be refinanced, and, to the extent relevant, the expected look-back period for refinanced ***green projects***. The GBP explicitly recognise several broad categories of eligibility for ***green projects***, which contribute to environmental objectives such as: climate change mitigation, climate change adaptation, natural resource conservation, biodiversity conservation and pollution prevention and control. The following list of project categories, while indicative, captures the most commonly used types of projects supported by or expected to be supported by the ***green bond*** market. ***Green projects*** include other related and supporting expenditures such as R&D and may relate to more than one category and/or environmental objective. Three environmental objectives identified above (pollution prevention and control, biodiversity conservation and climate change adaptation) also serve as project categories in the list. As such, they refer to the projects that are more specifically designed to meet them. The eligible ***green project*** categories, listed in no specific order, include, but are not limited to: ***Renewable energy*** (including production, transmission, appliances and products) ***Energy efficiency*** (such as in new and refurbished buildings, energy storage, district heating, smart grids, appliances and products) ***Pollution prevention and control*** (including reduction of air emissions, greenhouse gas control, soil remediation, waste prevention, waste reduction, waste recycling and energy/emission efficient waste-to-energy) ***Environmentally sustainable management of living natural resources and land use*** (including environmentally sustainable agriculture, environmentally sustainable animal husbandry, climate smart farm inputs such as biological crop protection or drip-irrigation, environmentally sustainable fishery and aquaculture, environmentally sustainable forestry, including afforestation or reforestation, and preservation or restoration of natural landscapes) ***Terrestrial and aquatic biodiversity conservation*** (including the protection of coastal, marine and watershed environments) ***Clean transportation*** (such as electric, hybrid, public, rail, non-motorised, multi-modal transportation, infrastructure for clean energy vehicles and reduction of harmful emissions) ***Sustainable water and wastewater management*** (including sustainable infrastructure for clean and/or drinking water, wastewater treatment, sustainable urban drainage systems and river training and other forms of flooding mitigation) ***Climate change adaptation*** (including information support systems, such as climate observation and early warning systems) ***Eco-efficient and/or circular economy adapted products, production technologies and processes*** (such as development and introduction of environmentally sustainable products, with an eco-label or environmental certification, resource-efficient packaging and distribution) ***Green buildings*** (meeting regional, national or internationally recognised standards or certifications)

(continued)

Table 2.3 (continued)

Component	*Guidelines*
Process for project evaluation and selection	The issuer of a ***green bond*** should clearly communicate to investors: The environmental sustainability objectives The process by which the issuer determines how the projects fit within the eligible ***green projects*** categories identified above The related eligibility criteria, including, if applicable, exclusion criteria or any other process applied to identify and manage potentially material environmental and social risks associated with the projects Issuers are encouraged to position this information within the context of the issuer's overarching objectives, strategy, policy and/or processes relating to environmental sustainability. Issuers are also encouraged to disclose any green standards or certifications referenced in project selection. The GBP encourage a high level of transparency and recommend that an issuer's process for project evaluation and selection be supplemented by an external review.
Management of proceeds	The net proceeds of the ***green bond***, or an amount equal to these net proceeds, should be credited to a sub-account, moved to a sub-portfolio or otherwise tracked by the issuer in an appropriate manner, and attested to by the issuer in a formal internal process linked to the issuer's lending and investment operations for ***green projects***. So long as the ***green bond*** is outstanding, the balance of the tracked net proceeds should be periodically adjusted to match allocations to eligible ***green projects*** made during that period. The issuer should make known to investors the intended types of temporary placement for the balance of unallocated net proceeds. The GBP encourage a high level of transparency and recommend that an issuer's management of proceeds be supplemented by the use of an auditor, or other third party, to verify the internal tracking method and the allocation of funds from the ***green bond*** proceeds.
Reporting	Issuers should make, and keep, readily available up-to-date information on the use of proceeds to be renewed annually until full allocation, and on a timely basis in case of material developments. The annual report should include a list of the projects to which ***green bond*** proceeds have been allocated, as well as a brief description of the projects and the amounts allocated, and their expected impact. Where confidentiality agreements, competitive considerations, or a large number of underlying projects limit the amount of detail that can be made available, the GBP recommend that information is presented in generic terms or on an aggregated portfolio basis (e.g. percentage allocated to certain project categories). Transparency is of particular value in communicating the expected impact of projects. The GBP recommend the use of qualitative performance indicators and, where feasible, quantitative performance measures (e.g. energy capacity, electricity generation, greenhouse gas emissions reduced/avoided, number of people provided with access to clean power, decrease in water use and reduction in the number of cars required), and disclosure of the key underlying methodology and/or assumptions used in the quantitative determination. Issuers with the ability to monitor achieved impacts are encouraged to include those in their regular reporting. Voluntary guidelines aiming at a harmonised framework for impact reporting exist for energy efficiency, renewable energy, water and wastewater projects, and waste management projects. The guidelines include templates for the format of impact reporting at a project and at a portfolio level that issuers can adapt to their own circumstances. The GBP encourage further initiatives, to help establish additional references for impact reporting that others can adopt and/or adapt to their needs. The use of a summary reflecting the main characteristics of a ***green bond*** or a ***green bond*** programme and illustrating its key features in alignment with the four core components of the GBP may help inform market participants.

Source: Adapted from ICMA (2018)

reported by issuers to stakeholders. Finally, the GBP recommend issuers, in connection with the issuance of a *green bond*, to appoint at least one external reviewer to confirm the alignment of their bond with the four core components of the GBP.[9] Even though the GBP are not mandatory standards, their development has played a significant role in structuring the *green bonds* market, providing all stakeholders with a first reliable tool able to effectively segregate *green bonds* from other debt securities. In this respect, certification agencies acting as reviewers (see also Sect. 2.3) today make wide reference to the GBP in their assessment activity, in this way prompting a certain degree of homogeneity in the market.

Nevertheless, it should be argued that the green debt market can hardly be considered to be limited to debt instruments formally in line with the GBP and eventually labelled as *green bonds*. As a matter of fact, a not negligible part of the unlabelled bonds outstanding could in principle meet the criteria set by the GBP, even though the issuers eventually disregarded the labelling option (e.g. in the case of many municipal bonds issued to finance projects of water pollution prevention). To define this category of debt issuances, the term *climate bonds* or *climate-aligned bonds* are often used. The size of this market, which is very difficult to calculate with accuracy, is indeed expected to be at least twice as large as the labelled *green bonds* market standalone (e.g. see CBI 2018b).

2.4 Definition of Green Loans

Following the introduction of the *Green Bond Principles*, a similar framework has been recently developed by the *Loan Market Association* (LMA) and the *Asian Pacific Loan Market Association* (APLMA), which in 2018 issued the *Green Loan Principles* (GLP). Largely inspired by the GBP, the GLP define *green loans* as "any type of loan instrument made available exclusively to finance or refinance, in whole or in part, new and/or existing eligible *green projects* [...] and are aligned with the four core components of the GLP" (LMA 2018). Guidelines for the four components of use of proceeds, process for project evaluation and selection, management of proceeds and reporting were developed accordingly. These guidelines are reported in Table 2.4.

[9] These external reviews can be of four types: second-party opinions, verifications, certifications or *green bond* scoring/ratings. For more details, see ICMA (2018).

Table 2.4 Core components of the green loan principles

Component	*Guidelines*
Use of proceeds	The fundamental determinant of a *green loan* is the utilisation of the loan proceeds for *green projects* (including other related and supporting expenditures, including R&D), which should be appropriately described in the finance documents. All designated *green projects* should provide clear environmental benefits, which will be assessed, and where feasible, quantified, measured and reported by the borrower. Where funds are to be used, in whole or part, for refinancing, it is recommended that borrowers provide an estimate of the share of financing versus refinancing. Where appropriate, they should also clarify which investments or project portfolios may be refinanced, and, to the extent relevant, the expected look-back period for refinanced *green projects*. A *green loan* may take the form of one or more tranches of a loan facility. In such cases, the green tranche(s) must be clearly designated, with proceeds of the green tranche(s) credited to a separate account or tracked by the borrower in an appropriate manner. The GLP explicitly recognise several broad categories of eligibility for *green projects* with the objective of addressing key areas of environmental concern such as climate change, natural resources depletion, loss of biodiversity, and air, water and soil pollution. To this extent, the list provided by the *Green Bond Principles* is taken as a reference. This non-exhaustive list is intended to capture the most usual types of projects supported, and expected to be supported, by the *green loan* market. However, it is recognised that definitions of green and *green projects* may vary depending on sector and geography.
Process for project evaluation and selection	The borrower of a *green loan* should clearly communicate to its lenders: The environmental sustainability objectives The process by which the borrower determines how the projects fit within the eligible *green projects* categories identified above The related eligibility criteria, including, if applicable, exclusion criteria or any other process applied to identify and manage potentially material environmental and social risks associated with the projects Borrowers are encouraged to position this information within the context of their overarching objectives, strategy, policy and/or processes relating to environmental sustainability. Borrowers are also encouraged to disclose any green standards or certifications to which they are seeking to conform.

Management of proceeds	The proceeds of a *green loan* should be credited to a dedicated account or otherwise tracked by the borrower in an appropriate manner, so as to maintain transparency and promote the integrity of the product. Where a *green loan* takes the form of one or more tranches of a loan facility, each green tranche(s) must be clearly designated, with proceeds of the green tranche(s) credited to a separate account or tracked by the borrower in an appropriate manner. Borrowers are encouraged to establish an internal governance process through which they can track the allocation of funds towards *green projects*.
Reporting	Borrowers should make and keep readily available up-to-date information on the use of proceeds to be renewed annually until fully drawn, and as necessary thereafter in the event of material developments. This should include a list of the *green projects* to which the *green loan* proceeds have been allocated and a brief description of the projects and the amounts allocated and their expected impact. Where confidentiality agreements, competitive considerations, or a large number of underlying projects limit the amount of detail that can be made available, the GLP recommend that information is presented in generic terms or on an aggregated project portfolio basis. Information need only be provided to those institutions participating in the loan. Transparency is of particular value in communicating the expected impact of projects. The GLP recommend the use of qualitative performance indicators and, where feasible, quantitative performance measures (energy capacity, electricity generation, greenhouse gas emissions reduced/avoided, etc.) and disclosure of the key underlying methodology and/or assumptions used in the quantitative determination. Borrowers with the ability to monitor achieved impacts are encouraged to include those in regular reports.

Source: Adapted from LMA (2018)

The aim of the GLP is to create a high-level framework of market standards, providing a consistent methodology for use across the *green loan* market, whilst allowing the loan product to retain its flexibility, and preserving the integrity of the *green loan* market while it develops (LMA 2018). The introduction of the GLP can be hence considered a structured tentative to further strengthen the *green finance* market by integrating the specificities of the loans as financing instrument. An external review to confirm the alignment with the four core components of the GLP is also recommended, even though with lesser emphasis when compared with GBP.[10]

Despite the many similarities, at least two key elements differentiate the GLP and the GBP strongly. A first one concerns the entity to which the guidelines are mainly directed: the issuer in case of the GBP and the borrower in case of the GLP. The approach used in the GLP factually limits the role of organisations that operationally issue the financial instrument (i.e. in case of loans, banks and other financial intermediaries) in the process of determining a *green loan*. Such an approach may indeed result in a little engagement of financial intermediaries in the market. The second differentiating element regards the potential coverage of the two set of principles, with respect to the theoretical market (namely, all the instruments eligible in terms of use of proceeds). If the GBP could in principle apply to all the eligible bond issuances, the potential application of the GLP is fundamentally restricted to loans of large size issued to big companies (and eventually syndicated). This is due to the expected relatively high operational costs to be faced by the borrowers to manage a *green loan*, which are only partially proportional to its size (in particular as concerns the management of proceeds and reporting requirements). In point of fact, the administrative costs linked to the management a *green loan* may be too high to be attractive for loans of limited size. Overall, the mentioned elements may cast some doubt on the possible wide adoption of the GLP in the near future.

[10] In this respect, the GLP recommend an external review only "when appropriate" and the review may be "partial, covering only certain aspects of a borrower's *green loan* or associated *green loan* framework". In addition, self-certification can also be an option (see LMA 2018). Such an approach is mainly due to the tentative to diminish the administrative burden for borrowers linked to the issuance of a *green loan*.

2.5 Compliancy with Industry Standards

With the rise of global awareness regarding the role of private actors in economic, social and environmental sustainability, many organisations have developed a specialization in reviewing and certifying whether the financial products and operations that are marketed as green or have an underlying environmental objective to meet specific criteria or policy requirements. These organisations are today an important component of the *green finance* landscape and have the role of promoting homogeneity and transparency in the market and, consequently, fostering investors' confidence. Given the variety of the possible securities, products or services featuring *green finance*, several types of organisations and services have progressively emerged.

2.5.1 External Reviews for Green Securities

In order to provide transparency on the use of proceeds for green securities (today this refers in particular to *green bonds* and, to a lesser extent, to *green loans*), several organisations have developed specific methodologies aimed at verifying the effective orientation of projects with positive environmental impact. To this extent, Table 2.5 summarises the main types of external reviews currently available in the market and mentions the key service providers. These reviews refer to both pre-issuance and post-issuance services and can take the specific form of *second-party opinions, third-party assurance reports* or *green ratings*.[11] In most of the cases, the alignment with the standards progressively being developed by the industry, such as the GBP and GLP, represents an essential part of the review process (as previously mentioned, an external review is in fact particularly important in the processes of labelling a *green bond* or a *green loan*).

2.5.2 Impact and Sustainability Reporting[12]

As corporate environmental and social responsibility grows more important, reporting on social and environmental performance to the different

[11] Examples of green rating are S&P's *Global Ratings Green Evaluation* and Moody's *Green Bonds Assessment*. S&P foresees five classes: GB1 (excellent), GB2 (very good), GB3 (good), GB4 (fair), GB5 (poor). Moody's considers four main classes (E1, E2, E3, E4) and an overall score out of 100.

[12] For a more detailed discussion on sustainability reporting, see Chap. 3.

Table 2.5 Main types of external reviews for green finance securities

Timing	*Type of review*	*Key features*	*Example of service providers*
Pre-issuance	Second-party opinion	An assessment of the issuer's *green bond* or *green loan* framework, analysing the "greenness" of eligible projects or assets. The assessment is also aimed at the attribution of a green label	Oekom, Sustainalytics, Vigeo, CICERO, CECEP Consulting
Pre-issuance	Third-party assurance report	Through an assurance report, statement of whether the green issuance is aligned with the *Green Bond Principles, Green Loan Principles* or other industry standards	Audit firms
Pre-issuance	*Green bond* rating	An assessment of a bond's alignment with the *Green Bond Principles* and the integrity of its green credentials. This usually result in attribution of a formal green rating (hence assigning a "*shade of green*") and implies the attribution of a green label	Rating agencies (Moody's, RAM Holdings, R&I, S&P Global Ratings)
Post-issuance	Second-party opinion or third-party assurance report	Periodical update of pre-issuance assessments or first assessment post-issuance	Oekom, Sustainalytics, Vigeo, CICERO, CECEP Consulting, audit firms, scientific experts

Source: Authors' elaboration

stakeholders of a company also start playing an increasingly relevant role. In this respect, a number of recognised impact and sustainability reporting standards exist[13] today, and several organisations offer services that facilitate companies to meet the requirements developed by these standards. In addition, verifiers exist providing assurance that a company's report about its impact and sustainability activity has been subject to proper scrutiny. To this extent, service providers are also active in specific subjects, such as the EU Emission Trading Scheme (ETS)[14] or company carbon footprint.[15] As a matter of fact, impact and sustainability reporting is expected to be one of the cornerstones of the *green finance* market (as well as of the overall *sustainable finance*) already in the near future, on which market efficiency and company credibility and relevant environmental benchmarking will be increasingly built. Reporting and disclosure requirement will also be parts to fostering the debate surrounding the definition of *green finance*.

[13] One of the most important examples of these reporting standards are the *Global Reporting Initiative* (GRI) *Sustainability Reporting*. These standards help companies to disclose their sustainability information by reporting their impact on the economy, the environment and society, but are also used to identify and manage related risks and find new opportunities. These Standards are composed of three universal Standards (foundation, general disclosures and management approach) that are applicable to all organisations and 33 topic-specific standards that organisations can select and use in economic, environmental and social series. See also (GRI 2016). Other examples of industry standards are the *AA1000AS* developed by *AccountAbility*, or the *ISAE3000* developed by the *International Federation of Accountants* (IFAC).

[14] Initiated during the Kyoto Protocol in 1997, the ETS is a European policy plan meant to reduce greenhouse gas emissions through a cap-and-trade system, a system where an overall cap representing the maximum amount of greenhouse gas (GHG) emissions is set for every firm and companies can buy and sell trade carbon permits in order to reach emissions targets. Under this legislation, companies have specific requirements, such as monitoring and reporting on emissions to the regulator, producing annual reports, and obtaining verifications by a certification body. Verifiers generally have both national and international competences. They review GHG emissions monitoring and reporting systems, verify the emissions baseline and the installation's annual emissions.

[15] Outside of the ETS, many carbon footprint products and services addressing corporate needs are provided by verifiers. Companies can determine the carbon footprint of a specific event, which might not only include direct carbon emissions, but also other types of emissions. Companies can choose to implement and develop a voluntary carbon footprint reporting which can participate in fostering a positive company image.

2.6 The Risk of Greenwashing

Along with the development of *green finance* standards and the growth in the environmental-related certifications and reviews, the discussion on whether corporations and private actors can use deceptive marketing strategies to promote their products and build an environmentally friendly image has also been emerging. This phenomenon, commonly known as *greenwashing*, is not specific to *green finance* and first appeared in the consumer goods industry. In this respect, literature has already observed that *greenwashing* can take many forms, ranging from changing the name of a product to induce the perception that it comes from a natural environment (when it does not) to launching marketing campaigns conducted by polluting industries to foster a green image (e.g. Schmuck et al. 2018; Parguel et al. 2015). The same literature has also noticed that given the increase in the awareness in the society regarding the potential environmental impacts of the products purchased, formal or informal labelling such as "green" (but also "eco-friendly" or "sustainable") are eventually getting increasingly popular and effective in driving market demand. Consequently, it has been observed that many products have been benefiting from a form of green advertising even though the environmental claims put forward did not present the real characteristics of the product (e.g. Baum 2012; Delmas and Burbano 2011).

The risk of *greenwashing* is progressively consolidating in the *green finance* market. As a matter of fact, the lack of universal definitions and standards amplifies such a risk as it opens to several possible interpretations of what green means in the financial markets. For this reason, as the market for sustainability-related certifications and reviews continues to develop, regulation on communication regarding the environmental impact of financial securities, products and services marketed as green should be also expected to become stricter.

References

Baum, L. M. (2012). It's not easy being green. Or is it? A content analysis of environmental claims in magazine advertisements from the United States and United Kingdom. *Environmental Communication, 6*(4), 423–440.

Climate Bonds Initiative (CBI). (2018a, September). *Climate bonds taxonomy*. London.

Climate Bonds Initiative (CBI). (2018b, September). *Bonds and climate change. The state of the market 2018*. London.

Delmas, M. A., & Burbano, V. C. (2011). The drivers of greenwashing. *California Management Review, 54*(1), 64–87.

European Investment Bank and Green Finance Committee of China Society for Finance and Banking (EIB and GFC). (2017, November). *The need for a common language in green finance.*

Global Reporting Initiative (GRI). (2016). *Introducing the GRI standards.* Amsterdam.

Höhne, N., Khosla, S., Fekete, H., & Gilbert, A. (2012). *Mapping of green finance delivered by IDFC members in 2011.* Cologne: Ecofys.

International Capital Market Association (ICMA). (2018, June). *Green bond principles. Voluntary process guidelines for issuing Green bonds.* Zurich.

International Finance Corporation (IFC). (2017). *Green finance. A bottom-up approach to track existing flows.* Washington, DC.

Loan Market Association (LMA). (2018). *Green loan principles.* London.

Multilateral Development Banks (MDB). (2015, September). *Joint report on multilateral development banks' climate finance.*

Parguel, B., Benoit-Moreau, F., & Russell, C. A. (2015). Can evoking nature in advertising mislead consumers? The power of *executional greenwashing. International Journal of Advertising, 34*(1), 107–134.

Schmuck, D., Matthes, J., & Naderer, B. (2018). Misleading consumers with green advertising? An affect–reason–involvement account of greenwashing effects in environmental advertising. *Journal of Advertising, 47*(2), 127–154.

Technical Expert Group on Sustainable Finance (TEG). (2019, June). *Taxonomy. Technical report.* Brussels.

United Nations Environment Programme (UNEP). (2016, September). *Definitions and concepts. Background note* (Inquiry working paper 16/13). New York.

CHAPTER 3

The Development of Green Finance by Sector

Olaf Weber and Amr ElAlfy

3.1 The Role of Multilateral Development Banks (MDB)

Multilateral development banks (MDB) are major international financial institutions and organizations, such as the World Bank (WB), the European Bank for Reconstruction and Development (EBRD), or the Asian Development Bank (ADB), and have an important but also ambivalent role when it comes to green financing. On the one hand, they finance activities that have detrimental effects on the environment, such as coal power plants. On the other hand, they provide guidelines for green financing and finance green projects, and contribute to address climate change through green bonds and other green financial investments. Financing both, however, might be inefficient, in the sense that MDB may end up financing projects that are harmful to the climate and then sponsor projects that help mitigate these negative impacts. This may be considered by many as a suboptimal use of financial capital with adverse effects on climate change and even economic development.

O. Weber (✉) • A. ElAlfy
University of Waterloo, Waterloo, ON, Canada

M. Migliorelli, P. Dessertine (eds.), *The Rise of Green Finance in Europe*, Palgrave Studies in Impact Finance,
https://doi.org/10.1007/978-3-030-22510-0_3

The World Bank is a major financier of fossil fuel projects. In 2010, it invested USD 4.4 billion in fossil fuel projects in the developing world. However, other MDB (such as the European Bank for Reconstruction and Development or the Asian Development Bank) keep investing billions in fossil fuel projects, such as coal power plants, even though some of them announced the intention to stop financing coal (Kynge and Hook 2018). Between 2006 and 2011, the EBRD increased its annual coal finance from EUD 82 million to USD 359 million. Another example is financing the 4000 megawatt Tata Mundra coal-fired power station in Gujarat, India, which received USD 450 million in financing from both the World Bank Group's International Finance Corporation (IFC) and the Asian Development Bank (Ghio 2015). To the same extent, the Asian Development Bank has been a major funder of coal-fired power plants globally. Between 1994 and 2012, the institution was the third largest public international financier of coal-fired power plants, investing USD 3.9 billion in 21 projects (Yang and Cui 2012). In addition, Yuan and Gallagher (2015) state that a third of all MDB's financing in Latin America and the Caribbean is not green. This significant amount of finance flows into extractive industries, the generation of fossil fuels, and conventional infrastructure projects that can increase global climate change, cause local environmental problems, and eventually adversely impact local communities and stakeholders. A part of these investments comes from Inter-American Development Bank that, on the other side, develops guidelines for managing environmental and social risks (Nolet et al. 2014) and is one of the leaders in green finance in the Americas.

Nevertheless, it should be recognized that MDB play a major role in green finance. According to the World Bank, climate financing by the world's six largest MDB increased to USD 35.2 billion in 2017, a 28% increase from the previous year (African Development Bank et al. 2018). The Asian Development Bank, for instance, analyzes fossil fuel subsidies in Asian countries, such as Indonesia and Thailand, with regard to their possible adverse effects. Furthermore, the social and environmental assessment guidelines of the World Bank and IFC already set global environmental and social standards (International Finance Corporation 2012) and are also the basis for industry voluntary codes of conduct such as the *Equator Principles* for project finance. Overall, MDB financing is a significant source of climate finance planned and indeed needed in the future (Westphal et al. 2015). In 2015, after China pledged to infuse USD 3.2 billion into a developing country fund for climate change, the Asian

Development Bank, the World Bank, and others began pledging major increases in climate finance as well. The World Bank pledged to increase climate finance to USD 29 billion (an increase by one third) by 2025 and the Inter-American Development Bank pledged to make climate finance 25–30 percent of total lending by that year (Yuan and Gallagher 2015).[1]

Among all financial institutions, the World Bank, which is indeed the leading source of international development funding (Rosen 2000), is best positioned to impose environmental and social responsibility on multilateral development banks and other international financial institutions and to provide environmental and social guidelines for all projects and investments. This is also true for IFC that developed the IFC standards for environmental and social sustainability as well as approaches to greenhouse gas (GHG) accounting and assessment (Performance Standard 3). Furthermore, IFC developed the *Cleaner Production Program* for assessing opportunities to implement energy efficiency processes and to reduce GHG emissions in IFC's portfolio. Finally, a significant share of green bonds and climate bonds are issued by MDB and they were among the first issuing green bonds and climate bonds at all. Hence, about 25 percent of green and climate bonds are issued by MDB (Climate Bonds Initiative 2018).

3.1.1 Future Steps for Greening the MDB

MDB already play a significant role in climate and renewable energy finance that will probably increase in the future because of a stronger

[1] Given the significant market failures involved in shifting investment into sustainable infrastructure in particular in the Caribbean and in Latin America, and the fact that the region is in the midst of an economic downturn, development banks are essential to filling a USD 260 billion dollar annual infrastructure gap and a USD 110 billion dollar annual gap in financing for climate change (Yuan and Gallagher 2015). MDB may play a significant role to achieve the Sustainable Development Goal (SDG) 7 on infrastructure and SDG 9 on energy by investments in sustainable infrastructure and renewable energy. In this respect, Bhattacharya et al. (2015) argue that development banks can play an essential role to help move nations and regions from "business as usual outcomes" to "sustainable infrastructure outcomes". Finally, MDB can help domestic financial institutions to integrate sustainability into their business by making financing dependent on the implementation of social and environmental sustainability guidelines for banks. IFC is already coordinating the development of financial sector sustainability regulations in some emerging countries and should continue to do so to support the sustainability case for the financial sector (Oyegunle and Weber 2015). For more details on the Sustainable Development Goals, see Chap. 1.

demand for green finance. MDB should take climate change issues and green economy into consideration in all their financing decisions. They should avoid to finance projects that are harmful for the climate on the one hand and invest in climate change mitigation and adaptation on the other hand. Instead, all project assessments should include environmental and social criteria. Financing cannot take place in silos anymore but has to integrate all economic, environmental, and social aspects in finance decisions. In addition to influencing financial regulators, MDB should continue to influence the financial sector's voluntary codes of conduct to enable them to have a stronger impact on the environmental and social performance of financed projects or other investments (Weber and Oni 2015). The IFC Performance Standards on Environmental and Social Sustainability (International Finance Corporation 2012), for instance, are an example of how an MDB can influence the financial industry through standards and guidelines. However, there should be less focus on "*doing no harm*" to "*do good*". Most of the MDB guidelines so far focus on reducing negative social and environmental impacts. More emphasis might be placed on financing activities with positive impacts on the environment, such as green technologies or green infrastructure. Sustainable finance means to take economic, social, and environmental issues equally into account and to avoid trade-offs.

3.2 Industrial Companies: A Focus on the Growing Importance of Environmental and Sustainability Reporting

Since the economic crisis of 2008, an increasing number of companies and industrial institutions have been disclosing annual reports describing their activities in addressing environmental issues. If one is to analyze the origins of the term "company", one should refer to the Latin phrase "*com panis*", which means "the sharing of bread" (Khodorkovsky 2008), which reflects that corporations' responsibility toward their stakeholders. This is not a new trend or concept in the business discourse. However, the definition and measurement criteria for social and environmental responsibility continue to be a subject of debate among academic, businesses, and civil actors. This debate stems from the nature of these reports, which address diverse stakeholders and accordingly vary in the structure, information provided, and quality.

Corporations have realized that reporting on environmental and social issues can help achieve long-term profitability through developing a positive corporate image, which should satisfy stockholders interests. Voluntarily reporting can help organizations mitigate future risks and implement systems that proactively prepare for mandatory government regulations, which can be costly to businesses. As a result, firms can sustain the flexibility of the decision making at their ends. In essence, self-regulated reporting should help a company achieve sustainability stewardship, which can save firms time and cost in case mandatory government regulations are put in place (Gunningham et al. 1998). Decision makers use the reports to leverage financial and non-financial performance. Reporting should also enhance the decision-making processes through benchmarking corporate performance of other organizations and sectors (Rikhardson et al. 2005). Sustainability reporting should help a company achieve operational efficiency through cost reduction or increased sales that result from enhanced corporate reputation (Schaltegger and Wagner 2006). Finally, effective reporting should help external stakeholders and investors understand a firm's vision, mission, and performance levels which should enhance a firm's goodwill (Global Reporting Initiative 2017).

Reporting on environmental and social performance is a key component of Corporate Social Responsibility (CSR) reporting, which is currently mandated by organizations' diverse stakeholders. Wood (1991) emphasizes the positive correlation between CSR reporting and corporate legitimacy (Melnyk et al. 2003) and reduces risks and costs (Weber et al. 2008a, b). Kurucz et al. (2008) analyze social and environmental as a "business case", where CSR is an investment that should result in positive economic and social returns. However, the relationship between corporate, social, and environmental performance and corporate financial performance has been controversial given the inconsistent and variant relationship between the two variables (Orlitzky 2008).

3.2.1 *Evolution of Environmental Reporting in Industrial Organizations*

Environmental reporting has a long history as an approach to help managers enhance their corporate image and achieve corporate sustainability. Corporate reporting started in the nineteenth century in the form of conventional financial reporting, where institutions disclose their financial performance data to internal and external stakeholders in the form of annual financial statements. Although accounting methods quantifies natural and human resources as cost elements within a firm's production system, insufficient attention has been paid to environmental and social issues (Houldin 2001). Reporting evolved to include a social dimension, which started in the late 1960s, where corporations reported to labor unions on their social performance (e.g., working conditions and compensations). Social reporting, unlike conventional reporting, focuses on qualitative and non-financial terms (Gray 2002). Further, environmental reporting started in the 1970s, where it was highly influenced by the *Brundtland Commission*'s agenda, which proposed "*long-term environmental strategies that can achieve effective sustainable development to the year 2000 and beyond*" (Brundtland 1987). Environmental accounting emerged in this milieu, where accountants started reporting to management and external stakeholders on firms' environmental performance and impacts (Schaltegger and Burritt 2006; KPMG 2003). However, environmental scholars have been cynical about the foundations of environmental accounting since the primary focus is profit generation rather than addressing ecological and social challenges (Gray and Bebbington 2000). There are technical issues in the environmental accounting that can be attributed to the complexity of our systems that cannot be monetized using the existing conventional financial accounting tools. These cases are evident when natural resources have a scarce social value to local communities or when environmental damage cannot be reversed (MacDonald 2010). In the late 1980s, firms in Europe and the United States of America started to disclose information on their emissions after the implementation of the *Toxic Release Inventory* (TRI) program. The program allowed several firms to map their environmental management programs and disclose robust information to their management and external stakeholders on their environmental performance. Another impetus for environmental reporting was led by the US Securities and Exchange Commission (SEC) when it required public firms to incorporate and disclose "*environmental exposures*" exceeding USD 100,000 in their yearly reports. The SEC initiative paved the road for many reporting initiatives afterward since organizations have recognized the importance of environmental reporting (Davis-Walling and Batterman 1997). Furthermore, the 1990s took a broader dimension after the *Brundtland Commission*'s definition of sustainability, where corporations attempted to achieve competitive advantage via environmental stewardship. The literature on balancing the economic, environmental, and social aspects of corporate responsibility also appeared in the 1990s as a response to the limitations of social and environmental accounting. "*Triple Bottom Line*" accounting, also known as TBL or 3BL, was introduced in 1994 by the British scholar John Elkington. The 3BL shifted corporate reporting, which was dominated by the financial bottom line to encompass social and environmental performance evaluation (Elkington 1998). Environmental reports changed from a narrative format to supplement financial information that are core to firms' financial performance. The reports also included regulatory and management information that address shareholders, community members, management, and others. Gray (2002, 2006) highlights that sustainability reporting has been treating the three pillars (economic, social, and environmental) in isolation whereas integration is needed to provide relevant and reliable information regarding corporate sustainability. The interrelation between the three domains as interacting systems should provide reliable and material information regarding sustainability performance as well as the risk associated with corporate activities. In fact, sustainability accounting has ongoing challenges to consider and quantify non-financial data and incorporate forward-looking information (ICAEW 2003). Owen and O'Dwyer (2008) are skeptical about contemporary sustainability accounting frameworks, which lack a robust integration and financial materiality, which is core to setting corporate strategies.

It is worth mentioning that the twenty-first century was highly influenced by sustainable development. This was evident in the World Business Council for Sustainable Development's (WBCSD) definition of CSR as "*the continuing commitment by business to behave ethically and contribute to economic development while improving the quality of life of the workforce and their families as well as of the local community and society at large*" (World Business Council for Sustainable Development 2008). With the introduction of the United Nation's Sustainable Development Goals (SDGs), which are 17 goals that shape the United Nation's view of sustainability until 2030, organizations have drafted their environmental reports to disclose information regarding their firms' roles in achieving the SDGs (Kaya 2016).[2] In essence, organizations have been addressing environmental and CSR reporting from a socio-economic lens that balances corporate profits, environmental concerns, and societal needs. The World Business Council for Sustainable Development (WBCSD) made several attempts to create reporting platforms that scale up business performance toward

[2] See also Chap. 1.

achieving the UN's SDGs (WBCSD 2017). Shaping corporate performance and reporting around the 17 goals can help provide robust guidelines for decision makers to contribute positively toward society and environment. Unlike Millennium Development Goals (MDGs), which were mainly state-centered, the 2015 SDGs shape a transformative shift in government and private sector cooperation. In this regard, the WBCSD introduced free "SDG compass" for businesses. The compass is a guideline available free on WBCSD's website to help companies understand the SDGs, align the firm's goals and operations with the 17 goals, and assure the integration of environmental reporting and corporate sustainability into corporate governance (SDG Compass 2017).

As concerns the chronological development of industrial sustainability progress, Nattrass and Altomare (1999) show how organizations in the 1970s responded in a reactive approach to newly implemented environmental regulations and standards. In the 1980s, organizations optimized the use of their resources in a way that optimizes cost efficiency. Organizations proactively incorporated environmental management systems in the 1990s to become more eco-efficient and achieve corporate legitimacy. Beginning in the 2000s, corporations started implementing integrated social and environmental reporting that aims at enhancing corporate accountability and sustainability.

3.2.2 Environmental Reporting Initiatives and Guidelines

Burritt and Schaltegger (2010) argue that the systemization of reporting frameworks is "*the first step in a methodological development process towards sustainability accounting providing useful and high-quality information*". Several institutions made numerous attempts to establish reporting guidelines that aim at harmonizing sustainability reporting. For example, quality standard certification is issued by the International Organization for Standardization (ISO), namely the ISO 9000, to measure corporate quality performance. Other ISO certifications have been focusing more on environmental issues such as ISO 14001, which measures firms' interaction with ecological resources, ISO 14063 for Environmental Communications, and ISO 26000, which provides guidance on firms' social responsibility (ISO 2017). The ISO standards have been widely adopted by corporations in different sectors as a positive response to internal and external stakeholders, who advocate for eco-efficient operational strategies (Clapp 1998). Likewise, AccountAbility 1000 (AA1000) and

SustainAbility are renowned guidelines for enhancing corporate sustainability performance and stakeholder engagement in corporate governance. However, both platforms have been criticized for lack of integrated reporting, which stems from their dependence on a triple bottom line approach in measuring sustainability (Freeman et al. 2010).

Furthermore, all previous initiatives have been trying to provide acceptable guidelines for reporting. Herzig and Schaltegger (2006) define a guideline as "*a non-binding guidance document based on practical experiences*". On the other hand, regulations are usually enforced by governing institutions to ensure systemized reporting. Three entities have promoted the integration and standardization of reporting as the pinnacle of reporting. The first is the International Integrated Reporting Council (IIRC), which is a coalition of NGOs, regulators, and companies that aim at establishing an integrated reporting framework across the global business. The second is the Sustainability Accounting Standards Board (SASB), which focuses on the materiality of sustainability accounting in a way that helps managers disclose useful information for investors as well as other stakeholders. The last is the Global Reporting Initiative (GRI), which has been the most accepted and adopted reporting guidelines by global corporations in the last ten years. In 2011, KPMG surveyed the world's largest 250 corporations. The survey's result shows that 95 percent of participating companies provide annual reports on their sustainability performance, of which 80 percent follow the GRI guidelines (KPMG 2011).

The GRI is an independent international organization that has made extensive efforts since 1997 to institutionalize sustainability reporting. GRI aims at helping businesses, governments, and institutions understand and communicate their impacts on global sustainability issues (Global Reporting Initiative 2017). Although SASB and IIRC provide better integrated and material reporting frameworks, the GRI initiative has been more successful in transforming niche individual corporate efforts in CSR reporting into a more standardized global trend. In essence, GRI has been adopted by the majority of global market leading companies for CSR reporting and continuous to be replicated across different sectors (Fifka 2012).

Additionally, there has been a significant collaboration between the GRI board, SASB, and the Climate Disclosure Project (CDP), which represents a global disclosure system that allows organizations to measure, manage, and report on their environmental performance. Since 2017, GRI and CDP have been collectively working on enhancing the quality of environmental reporting, when the two non-profit organizations signed a memorandum of understanding that aims at the systemization of companies' reporting on climate change and water data. Both organizations reach 6000 organizations that follow their guidelines to report on environmental performance (Global Reporting Initiative 2018). Table 3.1 provides a summary of the predominant reporting frameworks that are currently used by industrial organizations.

3.2.3 *Institutional Pressures and Environmental Reporting*

Moreover, cooperative dialogues and industry pressures can help develop reporting standards (Herzig and Schaltegger 2006). Organizations conform to rules in the market to sustain their operational legitimacy and enhance their image, which is the core of the institutional theory. The conceptual foundations of institutionalism aim at explaining the institutional order in a way that describes how and why institutions behave similarly across different organizations. Fernando and Lawrence (2014) emphasize the impact of institutional theory on developing resilient social structures. Institutional theory links organizational practices, which include environmental reporting, to values and norms of a society in which an organization operates where isomorphic changes can result from coercive, mimetic, and normative pressures (Powell and DiMaggio 1991). Briefly, coercive isomorphic change is mandated by supranational institutions and governments, as evident in the case of South Africa, where sustainability reporting is currently mandatory. In other words, all publicly traded companies on the Johannesburg Stock Exchange must integrate sustainability reporting with financial reporting (Dupont-Enzer 2014). Mimetic isomorphism occurs when corporations imitate one another to meet societal pressures and enhance their image. Imitation can also stem from an instrumental approach, as CSR reporting has been viewed as a tool that helps corporations achieve efficient and effective results on economic and socio-ecological levels (Porter and Kramer 2011). Finally, institutional change can be justified from a normative approach where inter-organizational professionals and networks bring change (Fifka 2012).

Several organizations have been successful in achieving environmental stewardship while sustaining positive financial growth. Through strategic corporate social responsibility, corporations formulate and articulate their values to ensure that they meet the expectations of their stakeholder. A good example of a company that has strategically invested in its corporate sustainability and environmental and social responsibility is the 3 M Company, which is an American multinational conglomerate based in

Table 3.1 Reporting frameworks

Standard	*Focus*	*Why report*	*Scoring*	*Who reports*
CDP	Primarily GHG emissions but has grown to address water and forestry issues as well.	CDP holds the largest repository of corporate GHG emissions and energy use data in the world and is backed by nearly 800 institutional investors representing more than USD 90 trillion in assets. Its transparent scoring methodology helps respondents understand the steps expected from them.	Companies receive two separate scores for disclosure and performance using 100-point scale. CDP recognizes top scoring companies in the Carbon Disclosure Leadership Index (CDLI).	Public and private companies, cities, government agencies, NGOs, supply chains.
Sustainability Accounting Standards Board (SASB)	Public companies only. Industry-specific issues deemed material to investors.	SASB standards enable comparison of peer performance and benchmarking within an industry.	No scoring system. Instead SASB is a standardized methodology for reporting sustainability performance through the form 10-K.	Corporations and companies.
ISO	Systemization and improvement of institutions' environmental management practices.	Quality assurance purposes.	No scoring system.	Public and private companies, and NGOS.
Integrated reporting (IR)	The IR framework provides guiding principles and content elements to assist companies (and other organizations) in the preparation and presentation of integrated reports.	Unlike the 3BL, integrating reporting focuses on the interaction between the economic, social, and environmental pillars.	No scoring system.	Public and private institutions.
Global reporting initiative (GRI)	Corporate social responsibility (CSR) with an equal weight on environmental. Social and governance factors. Heavy on stakeholder engagement to determine materiality.	GRI is the most renowned reporting platform. With the launch of the standards in 2018, GRI continues to be the oldest and widely respected reporting methodology globally.	Focus on economic, environmental, social, governance aspects of corporate performance.	Public and private companies, cities, government agencies, and NGOs.

Source: Author's elaboration

Maplewood, Minnesota. 3 M has global sales of USD 30.2 billion annually and employs 89,446 people worldwide and produces more than 55,000 products that are sold in about 200 countries (3 M 2017). 3 M has been a pioneer in acknowledging global challenges such as scarcity of natural resources, greenhouse gas emissions, and climate change. Therefore, 3 M management implemented the *Pollution Prevention Pays* (3P) strategy, which aims at lowering the consumption of water, energy, and material in the production process. The company communicated to its shareholders that the profitability will be impacted in the short term due to the initial investment costs; however, the company will harvest the economic, social, and environmental returns through collaboration with all stakeholders.

3 M invested in selecting responsible suppliers who can comply with providing environmentally friendly materials. The company also invested in closed loop fund that helps other organizations with their recycling initiatives. As a result of this comprehensive sustainability strategy that effectively engaged suppliers, employees, stockholders, customers, and local communities, 3 M was able to reduce the aggregate production costs and witnessed an increase in the corporation's goodwill as a result of the company's good environmental reputation (Weber 2014). The company was also able to prevent 2.1 million tons of pollutants and save USD 2.1 billion since the launch of the 3P strategy (3M 2017). Consumer retention rate has increased as a result of their satisfaction from high-quality and eco-friendly products that have lower prices, which stem from the reduction in raw material cost (3M 2014). Unfortunately, this sustainability success story is not a common case.

3.3 Greening of the Banking and Finance Sector

3.3.1 *Main Phases in Greening of the Financial Sector*

The greening of banks and other financial intermediaries gradually began in the 1980s. It was mainly driven by increasing energy prices and by the introduction of environmental laws and regulations. Consequently, the financial industry started with the greening of their operations to save costs for energy, waste, and material inputs, such as paper. Another motivation to go green was to be a model for clients. If banks could demonstrate that greening their business helps them to save costs, their borrowers or investees might follow their example and also save money by addressing environmental costs. This increases their financial liquidity and consequently, reduces risks for banks. Furthermore, it decreases the likelihood of environmental fines and reputational issues that also decreases risks for lenders and investors. At about the same time, environmental regulations based on the "*polluter pays principle*" created the responsibility for environmental impacts for all businesses. This also created risks for the financial sector as lenders and investors. Environmental costs for greening business and production processes as well as fines for environmental impacts created financial risks for businesses that also created risks for their lenders and investors. As a consequence, banks started to manage the risks mainly in commercial credit risk management (Weber et al. 2008a). They introduced criteria to assess the environmental and sustainability risks of their borrowers to avoid losses caused by environmental risks. As research has demonstrated, this approach helped to decrease credit risks (Weber et al. 2015).

After having established processes and tools to manage financial risks related to environmental issues, the financial industry focused on green investment opportunities. Mutual funds, indices, and other green investment vehicles have been issued. The first of these products addresses investment in green technologies. Later, socially responsible (SRI) or responsible investment (RI) used environmental, social, and governance criteria to analyze potential investments. Instead of only investing in green technologies or services, SRI invested in environmental leaders, excluded environmental laggards, or engaged with investees to push them into a more environmentally friendly direction. The best-known products and services that have been introduced during this area are the Dow Jones Sustainability Index, sustainability rating agencies, such as Sustainalytics, and investment funds such as the Ariel Fund. Also in the 1990s climate finance came up supported by the Kyoto Protocol (Labatt and White 2007). Another event that influenced the financial sector was the launch of the Kyoto Protocol on climate change mitigation. The financial sector engaged in financial products and services for carbon reduction, carbon offsets, and financed projects under the Kyoto Protocol mechanism.

Climate finance resurrected with COP 21 in Paris in 2015. Since, the global community achieved an agreement with regard to climate mitigation and adaptation, it became obvious that finance is needed to be able to achieve the climate goals. Climate bonds and green bonds become increasingly popular during this time. These bonds are issued by private or public issuers to finance activities that address climate change or other environmental impact, such as air and water pollution (Reichelt 2010; Weber and Saravade 2019). Climate change does not only offer financing opportunities for the financial industry but also bears risks. The Governor of the Bank of England Mark Carny was one of the first financial sector representatives who warned that the stability of the financial industry might be affected by climate change (Carney 2015). Direct physical risks caused by extreme weather events might impact financial sectors' operations, for instance, through the flooding of branches and IT facilities. These direct risks might also affect borrowers and investees and consequently expose the financial industry to risks. Furthermore, reputation risks might occur if banks finance clients that significantly contribute to climate change, such as coal power plants. Another type of risk is transition risks. These risks occur because of the transition to a low-carbon economy. Such a change in the structure of the economy, however, means that the financial industry has to adapt to these new structures, new types of businesses, and new types of risks.

Connected with transition risks is the risk of stranded assets. They appear because of the unexpected devaluation of assets because of the low-carbon technology diffusion as well as energy efficiency and climate policy measures (Mercure et al. 2018). Consequently, the value of assets of firms in the fossil fuel industry might decline and expose lenders, investees, and shareholders to financial risks. Recently, risk-adjusted returns of fossil fuel shares already underperformed those of other industries (Henriques and Sadorsky 2017; Hunt and Weber 2019), and financial industry portfolios exposed fossil fuels might be at risk.

3.3.2 Voluntary Codes of Conducts in the Financial Industry

The financial industry addresses green finance through a number of voluntary codes of conducts. One of the first is the *United Nations Environmental Program Finance Initiative* (UNEP FI)—founded in 1992—that originally tried to integrate environmental aspects into credit risk management and operations. Ten years later, the GRI's Financial Services Sector Supplement was created as the first effort to standardize environmental and sustainability reporting in the financial sector. Many institutions involved in the GRI Financial Sector Supplement have been also involved in UNEP FI. The *Equator Principles* (EP) and the *Principles for Responsible Investment* (PRI) codes of conducts for sub-groups of financial products and services, project finance, and institutional investment. The EPs have been launched by ten project financing institutions in 2003 based on IFC's performance standards of environmental and social sustainability (International Finance Corporation 2011, 2012). One of the reasons for the launch has been NGO pressure on project financiers to consider environmental and social aspects in their financing decisions (Weber and Acheta 2014). Currently, the EPs have 94 members. Their goal is to determine, assess, and to manage environmental and social risk in projects to guarantee a minimum standard for due diligence and monitoring to support responsible risk decision making (The Equator Principles 2013). Hence, they are not focusing on green finance but rather on the avoidance of environmental risks, a focus they have been often criticized for (Lawrence and Thomas 2004; Wright and Rwabizambuga 2006). Another critique of the *Equator Principles* is that they do not address climate change appropriately and still allow project finance for coal and coal power plants (Weber 2016a). Therefore, the question remains, whether the Equator Principles will help to increase the ratio of green finance in project finance.

A second major initiative for greening the financial industry is the Principles for Responsible Investing (PRI). PRI has more than 2200 members. The initiative addresses six main principles, such as (1) incorporating ESG issues in investment analysis and decision making, (2) to be active owners that incorporate ESG in their ownership policies and practices, (3) seeking appropriate disclosure on ESG issues by their investees, (4) to promote acceptance and implementation of the Principles within the investment industry, (5) to work together to enhance the effectiveness in implementing the Principles, and (6) to report on activities and progress toward implementing the Principles. PRI helps their members integrate sustainability with their financial decision making for investments and ownership practices. Recently, UNPRI introduced reporting and assessment standards (Weber 2018) to ensure that members follow the principles and to avoid freeriding (Richardson and Cragg 2010). Again, the principles rather address the integration of ESG with investment decisions, but it does not address increasing the ratio of green investments.

Two other initiatives, The Global Impact Investing Network (GIIN) and the Global Alliance for Banking on Values (GABV), pursue a different approach to green and sustainable banking. They mainly focus on increasing the positive social and environmental impact of the financial industry. The GABV, founded in 2009, consists of 55 banks, microfinance institutions, and credit unions globally.[3] According to GABV, these members advance positive change in the banking sector to make it more transparent, and to support economic, social, and environmental sustainability, as well as the real economy. Hence, GABV is less focused on mitigating financial risks caused by environmental issues but tries to use finance to deliver sustainable economic, social, and environmental development.[4] Though the banks in the network are very successful financially, most of the banks are relatively small and the total assets under management are just over USD 160 billion. To become a member of the association, financial institutions must fulfill certain criteria related to value-based banking. They have to use the triple bottom line approach at the core of their business model, and should be grounded in communities, serve the real economy, and enable new business models to meet the needs of communities and the real economy. Furthermore, they should strive for long-term relationships with clients to be able to understand their need and risks. Also, they should be self-sustaining and resilient to outside disruptions, such as financial crises. Finally, members should have a transparent and inclusive governance model (Weber 2018). With regard to green finance, members of the association finance projects and enterprises active in projects, such as clean energy, organic agriculture and food production, and zero waste projects.

The GIIN is an association addressing impact investing. Impact investing intentionally invests to generate positive environmental and social impacts (Weber 2016b). Conventional financial institutions conduct it as a part of their business, by philanthropists, and by specialized impact investors. The GIIN has developed the IRIS standards (www.thegiin.org/iris) for impact investment reporting. In contrast to UNEPFI and PRI, these standards measure the impact of the investment on the environment and society. The indicators can be selected based on the intended impact and address the categories presented in Table 3.2. The indicators are used by impact investors to assess the impact of their investments and to compare them with other investment or other investors. Furthermore, they can be used by stakeholders to evaluate investors.

[3] http://www.gabv.org/the-community/members/banks
[4] www.gabv.org/about-us

Table 3.2 Impact investing categories and indicators

Category	*Indicators*
Financial performance	Standard financial reporting metrics such as current assets and financial liabilities
Operational performance	Governance policies, employment practices, and social and environmental impact of day-to-day business activities
Product performance	Social and environmental benefits of the products, services, and unique processes offered by investees
Sector performance	Impact particularly in social and environmental sectors, including agriculture, financial services, and healthcare
Social and environmental objective performance	Progress toward specific impact objectives

Source: Author's elaboration

A similar direction with regard to impact is taken by the Principles for Responsible Banking. They are a part of UNEP FI and focus on addressing climate change and on creating a positive impact.[5] Forty-nine banks and a number of stakeholders have endorsed them as of March 2019. The principles state that banks align with the SDGs and the Paris Climate Goals. Furthermore, banks strive to work on achieving positive impacts through their business, and they work with their clients to encourage sustainable business practices. Fourthly, signatories proactively consult and engage stakeholders. Fifth, they will establish governance practices to achieve the targets, and finally, they are transparent and accountable for positive and negative impacts of their business (UNEP Finance Initiative 2018). With these principles, the UNEP Financial Initiative is the first "conventional" financial industry code of conduct that explicitly addresses the impact of banks on sustainable development and climate change. Hence, it uses a similar approach as GABV and GIIN. Prior to this, most voluntary codes of conducts rather addressed environmental risks for the financial industry. Furthermore, the principles strive to be transparent about both positive and negative impact. So far, sustainability reporting rather focused on positive impacts without being transparent about negative impacts (Weber 2016a).

[5] https://www.unepfi.org/banking/bankingprinciples/

3.4 Regulatory Approaches

In addition to voluntary codes of conduct, some national and international regulatory approaches exist. Internationally, the Task Force on Climate-Related Financial Disclosures (TCFD), initiated by the Financial Stability Board, has been developing standardized indicators to assess climate-related risks and opportunities (Task Force on Climate-related Financial Disclosures 2017). Also, the European Union published the report "Financing a sustainable European economy", which strives to develop a road map for sustainable finance in Europe (EU High Level Expert Group in Sustainable Finance 2018). Finally, we will discuss two major national policies to green the financial industry, the Chinese Green Credit Policy (China Banking Regulatory Commission 2012) and the Bangladeshi Environmental Risk Management Guidelines (ERM) and Green Banking Guidelines (Bangladesh Bank 2011).

The Task Force on Climate-related Financial Disclosures (TCFD) was established in 2015 to address the reporting problem on climate-related risks and opportunities, and the need for standardized reporting to ensure that the financial industry is able to evaluate and manage climate change–related risks (TCFD 2017). Because current disclosures lack information on the financial implications of climate-related aspects, the TCFD recommends that climate-related disclosure represents relevant information; is specific and complete; is clear, balanced, and understandable; is consistent over time; is comparable among companies within a sector, industry, or portfolio; is reliable, verifiable, and objective; and is provided on a timely basis (TCFD 2017). As a result of the provision of the above mentioned information, the financial industry should be enabled to manage climate-related risks that might affect their lending and investment portfolios (TCFD 2017). Consequently, the TCFD published industry-specific key performance indicators that can be integrated into lending and investment decisions.

Furthermore, to enable the financial industry to address climate-related risks accordingly, the TCFD recommends the development and the use of climate-related scenarios (TCFD 2017), and has developed implementation guidelines to implement effective climate risk management practices (TCFD 2017). These indicators and guidelines might be a first step in the standardization of climate-related risks assessment in the financial industry. However, to green the industry, strategies have to integrate the indicators with financial decision making.

The EU High Level Expert Group in Sustainable Finance published their report end of 2018 (EU High Level Expert Group in Sustainable Finance 2018). Priorities related to green financing identified by the expert group are to identify priority areas for climate finance.[6] Furthermore, the report addresses the short-termism of the financial industry that has already been addressed by Mark Carney, Governor of The Bank of England who called it the tragedy of the horizon (Carney 2015). Another important recommendation of the report is to develop standards for green financial products and services, such as green bonds, to increase the transparency in the field. Also, the report recommends to integrate sustainability in both the governance of financial institutions and financial supervision.

The Chinese Green Credit Policy requires lenders to allocate investment toward green industries, to constrain investments in polluting industries, and to withdraw financing from industries targeted for their negative environmental impact (Weber 2017). State Environmental Protection Administration (SEPA), the People's Bank of China (PBOC), and the China Banking Regulatory Commission (CBRC) have published this policy. Banks have to deliver key performance indicators to the financial regulator who will use them for their risk assessment. Consequently, this is the first policy that implements financial sector sustainability regulations overseen by the financial regulators. Though implementation issues with regard to the policy are discussed controversially (Zhang et al. 2011), studies suggest a positive impact on both the increase of green lending and the decrease of financial risks (Cui et al. 2018). A longer-term evaluation will show whether the policy achieved its intended goal.

Another country that implemented green finance regulations through is central banking authority is Bangladesh. In 2011, they introduced the Environment Risk Management Guidelines (ERM) and Green Banking Guidelines in 2011 (Bangladesh Bank 2011). Since then, the policies have been upgraded by integrating environment and social risk with the Credit Risk Management (CRM) guidelines (Weber et al. 2015). Furthermore, Bangladesh Bank introduced Environmental and Social Risk Management (ESRM) guidelines including an environmental risk analysis model (Chowdhury 2018). Studies suggest that the introduction of environmental issues in credit risk analysis increases the quality of the risk rating process, because adding environmental and social aspects in the analyses

[6] For more information on the initiatives in the EU to foster sustainable finance, see Chap. 6.

increases the risk rating ability (Weber et al. 2015). However, other studies demonstrate that Bangladeshi banks adopt the policy because it is mandatory and consequently increase their financial performance. On the other side, however, they do not adopt sustainability practices on a voluntary basis because they want to benefit from this win-win-situation (Chowdhury 2018). Hence, it is important not only to introduce regulations and guidelines but also to educate the financial industry about the benefits of adopting a green finance strategy.

In general, green and sustainability guidelines and regulations overseen by financial regulators are in their infancy. First results seem to be positive with regard to decreasing financial risks and increasing green finance. However, more research is needed to explore longer-term effects and the effectiveness of different regulations in different countries and regions.

3.5 Conclusions

This chapter reported about approaches in green finance by multilateral financial institutions, industrial companies, and banks. In all these three sectors green finance is on the rise, be it to reduce costs by reducing the use of energy and other resources as well as mitigating risks, or be it to increase revenues by offering green finance and green finance products and services. Hence, financial materiality seems to be the main driver for green finance so far. Though we see an increase in green finance, we also have to conclude that green finance is far from being in the core of the business for most MDB, industrial companies, and banks. For most of them green finance is a niche product and service compared to their conventional business. MDB financing green energy and coal at the same time, fossil fuel companies that also invest in renewable energy, and banks that lend to the oils sands and green tech at the same time are the rule and not the exception. This might make sense from a portfolio diversification perspective. However, it does not make sense from a longer-term impact perspective because negative impacts of conventional finance might materialize for financial institutions and companies in the future. For instance, increased extreme weather events, resulting from emissions and financed emissions will have a negative effect on the economy and its players.

If we have a look on reporting, one might get the impression that green finance plays a major role in MDB, companies, and banks. This, however, is less a matter of the ratio of green finance compared to other businesses, but it is because of the way of reporting. Most of the reporting is still to paint a positive picture to stakeholders and shareholders. It is used less as a strategic management tool, but as a tool to increase the reputation of firms.

Furthermore, many of the reporting standards focus on what is profitable for the company and not for the environment. Consequently, performance is reported from an investor's perspective. It is less about the impact of green finance on the environment, but rather about the impact of green finance on the company itself. This supports green finance only as far as it has a direct positive impact on the business or as long as it has a positive impact on reputation. Environmental reporting and accounting, however, should also account for the positive and negative impacts of green and conventional finance on the environment. Therefore, to create a transparent picture of green finance, both green and brown finance have to be reported. Hence, to conclude this section, we state that green finance is on the rise. However, it is still reactionary instead of being a strategic core business and a holistic approach that weighs green finance against brown finance, which is still missing.

References

3M. (2014, July 1). *Environment, health, and safety standard.* Retrieved from: http://multimedia.3m.com/mws/media/731894O/waste-management-program-standard.pdf?&fn=WasteMgmntProgramStandard.pdf

3M. (2017). *2016 Sustainability report.* Retrieved from: http://www.3m.com/3M/en_US/sustainability-report/global-challenges/raw-materials/

African Development Bank, Asian Development Bank, European Bank for Reconstruction and Development, European Investment Bank, Inter-American Development Bank Group, Islamic Development Bank, and World Bank Group. (2018). *Joint report on multilateral development banks' climate finance.* London: European Bank for Reconstruction and Development.

Bangladesh Bank. (2011). *Environmental risk management (ERM) guidelines for banks and financial institutions in Bangladesh.* Retrieved from Dhaka, Bangladesh.

Bhattacharya, A., Oppenheim, J., & Stern, N. (2015). *Driving sustainable development through better infrastructure: Key elements of a transformation program.* Retrieved from Washington, DC.

Brundtland, G. H. (1987). *Our common future.* Oxford: Oxford University Press.

Burritt, R. L., & Schaltegger, S. (2010). Sustainability accounting and reporting: Fad or trend? *Accounting, Auditing & Accountability Journal, 23*(7), 829–846.

Carney, M. (2015). *Breaking the tragedy of the horizon – Climate change and financial stability*. London: Bank of England.

China Banking Regulatory Commission. (2012). *Notice of the China banking regulatory commission CBRC on issuing the green credit guidelines*. Beijing: China Banking Regulatory Commission.

Chowdhury, R. K. (2018). *Corporate sustainability and financial performance of Bangladeshi banks*. Waterloo: University of Waterloo. Retrieved from: http://hdl.handle.net/10012/14208.

Clapp, J. (1998). The privatization of global environmental governance: ISO 14000 and the developing world. *Global Governance, 4*(3), 295–316.

Cui, Y., Geobey, S., Weber, O., & Lin, H. (2018). The impact of green lending on credit risk in China. *Sustainability, 10*(6), 2008. https://doi.org/10.3390/su10062008.

Davis-Walling, P., & Batterman, S. A. (1997). Environmental reporting by the fortune 50 firms. *Environmental Management, 21*(6), 865–875.

Dupont-Enzer, M. (2014, April 2). *Why is South Africa leading the integrated reporting debate?* Retrieved from: http://publisher.enablon.com/blog/why-is-south-africa-leading-the-integrated-reporting-debate/

Elkington, J. (1998). *Cannibals with forks*. London: Chapman.

EU High Level Expert Group in Sustainable Finance. (2018). *Financing a sustainable European economy*. Brussels: European Commission.

Fernando, S., & Lawrence, S. (2014). A theoretical framework for CSR practices: Integrating legitimacy theory, stakeholder theory and institutional theory. *The Journal of Theoretical Accounting Research, 10*(1), 149–178.

Fifka, M. (2012). The development and state of research on social and environmental reporting in global comparison. *Journal Für Betriebswirtschaft, 62*(1), 45–84.

Freeman, R. E., Harrison, J. S., Wicks, A. C., Parmar, B. L., & De Colle, S. (2010). *Stakeholder theory: The state of the art*. Cambridge: Cambridge University Press.

Ghio, N. (2015, April 21). *New report once again shows danger of MDB financing for large fossil fuel projects*. Retrieved from: http://www.huffingtonpost.com/nicole-ghio/new-report-once-again-sho_b_7102638.html

Global Reporting Initiative. (2018). Retrieved from: http://database.globalreporting.org/

Gray, R. (2002). The social accounting project and accounting organizations and society privileging engagement, imaginings, new accountings and pragmatism over critique? *Accounting, Organizations and Society, 27*, 687–708.

Gray, R. (2006). Social, environmental and sustainability reporting and organisational value creation? whose value? whose creation? *Accounting, Auditing & Accountability Journal, 19*(6), 793–819. https://doi.org/10.1108/09513570610709872.

Gray, R., & Bebbington, J. (2000). Environmental accounting, managerialism and sustainability: Is the planet safe in the hands of business and accounting? *Advances in Environmental Accounting and Management, 1*, 1–44.

GRI and SASB take next step in cooperation. (2017, May 4). Retrieved from: https://www.globalreporting.org/information/news-and-press-center/Pages/GRI-and-SASB-take-next-steps-in-cooperation.aspx

Gunningham, N., Grabosky, P., & Sinclair, D. (1998). *Smart regulation: Designing environmental policy*. Oxford: Oxford University Press.

Henriques, I., & Sadorsky, P. (2017). Investor implications of divesting from fossil fuels. *Global Finance Journal*. https://doi.org/10.1016/j.gfj.2017.10.004.

Herzig, C., & Schaltegger, S. (2006). Corporate sustainability reporting: An overview. In S. Schaltegger, M. Bennett, & R. Burritt (Eds.), *Sustainability accounting and reporting* (pp. 301–324). Dordrecht: Springer.

Houldin, M. (2001). Overview. In R. Gray, J. Bebbington, & D. Walters (Eds.), *Accounting for the environment*. London: Chapman Publishing.

Hunt, C., & Weber, O. (2019). Fossil fuel divestment strategies: Financial and carbon related consequences. *Organization and Environment, 32*(1), 41–61. https://doi.org/10.1177/1086026618773985.

Initiative, C. B. (2018). *Bonds and climate change – The state of the market 2018*. London: Climate Bonds Initiative.

Institute of Chartered Accountants in England Wales. (2003). *Information for better markets new reporting models for business*. London: ICAEW.

International Finance Corporation. (2011). *Update of IFC's policy and performance standards on environmental and social sustainability, and access to information policy*. Retrieved from Washington, DC.

International Finance Corporation. (2012). *IFC performance standards on environmental and social sustainability* (p. 72). Washington, DC: International Finance Corporation. http://www.ifc.org/wps/wcm/connect/topics_ext_content/ifc_external_corporate_site/ifc+sustainability/publications/publications_handbook_pps.

ISO. (2017). *International organization for standardization*. Retrieved from: https://www.iso.org/home.html

Kaya, I. (2016). The mandatory social and environmental reporting: Evidence from France. *Social and Behavioral Sciences, 229*, 206–213.

Khodorkovsky, M. (2008, September 29). *Etymology of company*. Retrieved from: https://www.altalang.com/beyond-words/2008/09/29/etymology-of-companion/

KPMG. (2003). *Building a new reporting and communications model: A new source of competitive advantage*. Melbourne: KPMG.

KPMG. (2011). *International survey of corporate responsibility reporting 2011*. Amsterdam.

Kurucz, E. C., Colbert, B. A., & Wheeler, D. (2008). The business case for corporate social responsibility. In A. Crane, A. McWilliams, D. Matten, & J. Moon (Eds.), *The Oxford handbook of corporate social responsibility* (1st ed., pp. 83–112). Oxford: Oxford University Press.

Kynge, J., & Hook, L. (2018, December 12). Development bank halts coal financing to combat climate change. *Financial Times*.

Labatt, S., & White, R. R. (2007). *Carbon finance*. Hoboken: Wiley.

Lawrence, R. F., & Thomas, W. L. (2004). The equator principles and project finance: Sustainability in practice? *Natural Resources and Environment, 19*(2), 20.

MacDonald, K. I. (2010). The devil is in the (bio)diversity: Private sector engagement and the "restructuring" of biodiversity conservation. *Antipode, 42*(3), 513–550.

Melnyk, S. A., Sroufe, R. P., & Calantone, R. (2003). Assessing the impact of environmental management systems on corporate and environmental performance. *Journal of Operations Management, 21*, 329–351.

Mercure, J. F., Pollitt, H., Viñuales, J. E., Edwards, N. R., Holden, P. B., Chewpreecha, U., et al. (2018). Macroeconomic impact of stranded fossil fuel assets. *Nature Climate Change*. Retrieved from doi:https://doi.org/10.1038/s41558-018-0182-1.

Nattrass, B., & Altomare, M. (1999). *The natural step for business*. Gabriola Island: New Society Publishers.

Nolet, G., Vosmer, W., de Bruijn, M., & Braly-Cartillier, I. (2014). *Managing environmental and social risks, a roadmap for National Development Banks in Latin America and the Caribbean*. Retrieved from Washington, DC. http://publications.iadb.org/bitstream/handle/11319/6437/CMF%20MON%20Managing%20Environmental%20and%20Social%20Risks.pdf?sequence=1

Orlitzky, M. (2008). Corporate social performance and financial performance: A research synthesis. In A. Crane, D. Matten, A. McWilliams, J. Moon, & D. S. Siegel (Eds.), *The Oxford handbook of corporate social responsibility*. New York: Oxford University Press on Demand.

Owen, D. L., & O'Dwyer, B. (2008). Corporate social responsibility: The reporting assurance dimension. In A. Crane, D. Matten, A. McWilliams, J. Moon, & D. S. Siegel (Eds.), *The Oxford handbook of corporate social responsibility*. Oxford: Oxford University Press on Demand.

Oyegunle, A., & Weber, O. (2015). *Development of sustainability and green banking regulations – Existing codes and practices (65)*. Retrieved from Waterloo. https://www.cigionline.org/publications/development-of-sustainability-and-green-banking-regulations-existing-codes-and-practice

Porter, M. E., & Kramer, M. R. (2011). Creating shared value: How to reinvent capitalism – And unleash a wave of innovation and growth. *Harvard Business Review, 89*(1), 2.

Powell, W., & DiMaggio, P. (1991). *The new institutionalism in organizational analysis*. Chicago: University of Chicago Press.

Reichelt, H. (2010). Green bonds: A model to mobilise private capital to fund climate change mitigation and adaptation projects. *The EuroMoney Environmental Finance Handbook*, 1–7.

Richardson, B. J., & Cragg, W. (2010). Being virtuous and prosperous: SRI's conflicting goals. *Journal of Business Ethics, 92*, 21–39. https://doi.org/10.1007/s10551-010-0632-9.

Rikhardson, P., Bennett, M., Bouma, J., & Schaltegger, S. (2005). Environmental management accounting: Innovation or managerial fad? In P. Rikhardsson, M. Bennett, J. Bouma, & S. Schaltegger (Eds.), *Implementing environmental management accounting. Status and challenges* (pp. 1–18). Dordrecht: Springer.

Rosen, C. (Ed.). (2000). *World resources 2000–2001: People and ecosystems: The fraying web of life*. Amsterdam: Elsevier.

Schaltegger, S., & Burritt, R. (2006). Corporate sustainability accounting: A nightmare or a dream coming true? *Business Strategy and the Environment, 15*, 293–295.

Schaltegger, S., & Wagner, M. (2006). Integrative management of sustainability performance, measurement and reporting. *International Journal of Accounting, Auditing and Performance Evaluation, 3*(1), 1–19.

SDG Compass. (2017). *The goals*. Retrieved from: https://sdgcompass.org/sdgs/

Task Force on Climate-related Financial Disclosures. (2017). *Recommendations of the task force on climate-related financial disclosures*. Basel: Task Force on Climate-related Financial Disclosures.

The Equator Principles. (2013). *The equator principles*. Retrieved from Geneva.

UNEP Finance Initiative. (2018). *Principles for responsible banking – Shaping our future*. Geneva: UNEP Finance Initiative.

WBCSD. (2017). *World business council for sustainable development*. Retrieved from: http://www.wbcsd.org/

Weber, O. (2014). Environmental, social and governance reporting in China. *Business Strategy and the Environment, 23*(5), 303–317. https://doi.org/10.1002/bse.1785.

Weber, O. (2016a). Equator principles reporting: Factors influencing the quality of reports. *International Journal of Corporate Strategy and Social Responsibility, 1*(2), 141–160.

Weber, O. (2016b). Impact investing. In O. M. Lehner (Ed.), *Routledge handbook on social and sustainable finance* (pp. 85–101). London: Routledge.

Weber, O. (2017). Corporate sustainability and financial performance of Chinese banks. *Sustainability Accounting, Management and Policy Journal, 8*(3), 358–385. https://doi.org/10.1108/SAMPJ-09-2016-0066.

Weber, O. (2018). Financial sector sustainability regulations and voluntary codes of conduct: Do they help to create a more sustainable financial system? In T. Walker, S. D. Kibsey, & R. Crichton (Eds.), *Designing a sustainable financial system: Development goals and socio-ecological responsibility* (pp. 383–404). Cham: Springer International Publishing.

Weber, O., & Acheta, E. (2014). *The equator principles: Ten teenage years of implementation and a search for outcome* (CIGI papers series) (p. 20). Waterloo: CIGI.

Weber, O., & Oni, O. (2015). *The impact of financial sector sustainability regulations on banks (65)*. Retrieved from Waterloo, ON.

Weber, O., & Saravade, V. (2019). *Green bonds – Current development and their future* (p. 32). Waterloo: Centre for International Governance Innovation.

Weber, O., Fenchel, M., & Scholz, R. W. (2008a). Empirical analysis of the integration of environmental risks into the credit risk management process of European banks. *Business Strategy and the Environment, 17*, 149–159. https://doi.org/10.1002/bse.507.

Weber, O., Koellner, T., Habegger, D., Steffensen, H., & Ohnemus, P. (2008b). The relation between the GRI indicators and the financial performance of firms. *Progress in Industrial Ecology, 5*(3), 236–254.

Weber, O., Hoque, A., & Islam, A. M. (2015). Incorporating environmental criteria into credit risk management in Bangladeshi banks. *Journal of Sustainable Finance and Investment, 5*(1–2), 1–15. Retrieved from https://doi.org/10.1080/20430795.2015.1008736.

Westphal, M. I., Canfin, P., Ballesteros, A., & Morgan, J. (2015). *Getting to $100 billion: Climate finance scenarios and projections to 2020.* Retrieved from Washington, DC: http://www.wri.org/publication/getting-100-billion-climate-finance-scenarios-and-projections-2020

Wood, D. J. (1991). Corporate social performance revisited. *Academy of Management Review, 16*(4), 691–718.

Wright, C., & Rwabizambuga, A. (2006). Institutional pressures, corporate reputation, and voluntary codes of conduct: An examination of the equator principles. *Business and Society Review, 111*(1), 89–117. Retrieved from https://doi.org/10.1111/j.1467-8594.2006.00263.x.

Yang, A., & Cui, Y. (2012). *Global coal risk assessment: Data analysis and market research.* Retrieved from Washington, DC.

Yuan, F., & Gallagher, K. P. (2015). *Greening development finance in the Americas.* Retrieved from Boston, MA.

Zhang, B., Yang, Y., & Bi, J. (2011). Tracking the implementation of green credit policy in China: Top-down perspective and bottom-up reform. *Journal of Environmental Management, 92*(4), 1321–1327. Retrieved from: http://www.sciencedirect.com/science/article/pii/S0301479710004639. https://doi.org/10.1016/j.jenvman.2010.12.019.

CHAPTER 4

Sustainable Finance Management

Giovanni Ferri and Francesca Lipari

4.1 Introduction

The unfolding of the Green Finance paradigm represents undoubtedly one of the most significant evolutions in the financial landscape. If only we go back to the major responsibilities of finance in bringing the world economy close to collapse with the Global Financial Crisis (GFC) of 2007–2009, one must admit that something has changed. While exploitative and profit-seeking behaviour may be still engulfing part of the financial industry, a new brand of ethical and sustainable finance seems to be emerging ten years after the GFC, as documented in a large part of this book.

The issue we aim to address in this chapter is how firms need to shape up or reshape in order to become suitable for the Green Finance evolution. Specifically, the main conclusion we reach is that, to tap Green Finance, a firm has to structure—if a start-up—or reorganise—if already in business—in a way to be sustainability compliant and, what's more, to be able to display that credibly. In turn, this passage demands that a firm adopts a sustainable finance management approach, which may be achieved by hiring sustainable finance managers—especially in medium–

G. Ferri • F. Lipari (✉)
LUMSA University, Rome, Italy
e-mail: g.ferri@lumsa.it; f.lipari@lumsa.it

M. Migliorelli, P. Dessertine (eds.), *The Rise of Green Finance in Europe*, Palgrave Studies in Impact Finance,
https://doi.org/10.1007/978-3-030-22510-0_4

large firms—or outsourcing that function to external consulting experts. In both cases, the firm needs to be able to get a good external evaluation, typically a credible ESG (Environment, Social, Governance) rating. Socially Responsible Investment (SRI) funds, Impact investors and sustainability-oriented asset managers more in general all typically employ ESG ratings as their rule of thumb or compass.

In this chapter, we reach the above conclusions based on the following reasoning. In Sect. 4.2, we detail the financial needs of a firm subscribing a sustainable approach. In particular, we distinguish between sustainable start-ups (Sect. 4.2.1)—which are entirely free to structure themselves as sustainable—and existing firms in transition to a sustainable approach (Sect. 4.2.2)—which face instead the challenge of letting a new sustainable business emerge, possibly as a carve out inside the existing concerns. In Sect. 4.3, we go in-depth about the managerial and organisational implications for accessing Green Finance. Specifically, we highlight that, under the guidance of a sustainable finance manager, a sustainable start-up or an existing firm in transition to a sustainable approach can benefit from adopting specific organisational set ups. Section 4.4 discusses a preliminary empirical check of the nexus between organisational structures and ESG ratings. Namely, referring to the 38 listed firms included in the FTSE MIB30 basket for which the appropriate information could be retrieved, we document that, indeed, organisational structures in which sustainability is more credibly systematised—and gains more centrality within the firm's organisational chart—seem to obtain higher ESG ratings. Finally, Sect. 4.5 concludes recapitulating the main thrust of our argument and discussing potential policy implications.

4.2 Financial Needs of Firms Subscribing a Sustainable Approach

The road towards green growth and sustainability is tortuous and often features disruptive events. It requires a new mindset for imagining a product or service, or for using resources (both natural and human resources) and for doing business in different ways than usually done. In a nutshell, a sustainable approach to growth entails innovation.

Fichter (2005) defines sustainable innovation as "the development and implementation of a radically new or significantly improved technical, organisational, business-related, institutional or social solution that meets a triple bottom line of economic, environmental and social value creation. Sustainable innovation contributes to production and consumption patterns that secure human activity within the earth's carrying capacities" (Fichter 2005, p. 138, authors' translation). In this chapter, we will adopt this concept of "sustainable innovation". Examples of existing sustainable innovation include organic and fair food production, electric and shared mobility, sustainable fashion, renewable energy technology, energy-efficient "smart homes" and eco-tourism, though this list is far from exhaustive.

Innovation and the related process of creative destruction will lead to new ideas, new entrepreneurs and new business models, thus contributing to the establishment of new markets and eventually to the creation of new jobs. Green innovation is therefore the key in enabling environmentally sustainable growth. But how?

According to Lazonick (2015), there are specific features of innovation in general that are important for sustainable firms. An innovative firm is one that has implemented an innovation during the productive transformation. Such innovation is supported by:

- *Strategy*: The innovation process is *uncertain* because what needs to be learned about transforming technologies and accessing markets can only become known through the process itself hence, given that optimisation is not viable, we need strategic control to allocate resources to investments in developing human and physical capabilities that will enable the firm to compete.
- *Organisation*: The innovation process is a *collective* one because it requires the collaboration of different people with different capabilities within and outside the firm; hence, we need organisational integration to learn how to transform technologies and access markets in ways that generate higher quality and/or lower costs.
- *Finance*: The innovation process is *cumulative* through time when learning cannot be done all at once: what is learned today provides a foundation for what can be learned tomorrow, and these organisational learning process must be sustained over time. Such support is provided through financial commitment that sustains the process of developing technologies, recalibrating their internal organisation and accessing markets from the strategy steps until financial returns are generated.

Innovation is a crucial dimension for measuring the capacity of a business to place itself on the production frontier and to gather the fruits of being the "first mover". But this first mover advantage does not come for free. It highly relies on financial support.

Finance, indeed, plays a critical role in innovation as it allows organisations to conduct activities of R&D, adopt technologies and internal structure necessary for sustainable inventions as well as develop and commercialise such innovations.

Firms can use either internal or external sources of finance to fund their innovation activities.

The main internal source of finance is retained earnings, the profits accumulated over time which have not been returned to shareholders. Firms typically prefer to use internal financing rather than external financing as the latter can be very costly. But there is a further reason why innovative firms may have to rely disproportionately on internal-raised funds. Indeed, innovations typically twist the balance within the firm against tangible assets and in favour of intangible assets. In turn, since pledging intangible assets is harder than pledging tangible assets (Hart and Moore 1994; Brown et al. 2009; Hall and Lerner 2010), innovative firms are more likely to experience rationing of external funds.

As a result, there are projects that a firm would choose to undertake if it had sufficient internal resources available, but which will be shelved if the firm needs to access external finance to develop them. In many cases, as said, a firm does not have the option to access external financing. However, in many others, especially for sustainable innovation, they need to resort to external funding.

External sources of financing include debt and equity (as well as some hybrid forms), which can be provided by individual investors (such as business angels), venture capital funds, banks and capital markets (among others). Conditional on having to resort to external funds, debt is generally preferred to equity, since, if available, debt is typically a cheaper source of finance (even if still more expensive than internal funds). For example, large firms can more easily finance their R&D activities, whether using internal resources, getting a loan from a bank (using their tangible assets as collateral if required), issuing bonds, or raising equity finance in the stock markets. Start-ups do not have as many assets to use as collateral and their innovation investment is less diversified and may also represent a much larger share of their activities for really innovative firms. As a result, their funding options are much more limited.

The access to external finance for innovation is an important challenge for firms. The challenges, especially in the realm of Sustainable Innovation are basically two: supply of funding and access to funding.

Concerning the first, firms can fund innovation activities using a variety of funding instruments provided by different types of financial intermediaries and investors. Specifically, Sustainable financing aims to increase the level of financial flows (from banking, micro-credit, insurance and investment) from the public, private and not-for-profit sectors to SDGs challenges and priorities. Sustainable finance refers to any form of financial service integrating environmental, social or governance (ESG) criteria in business or investment decisions, for the lasting benefit of both clients and society at large. A key part of this is to be oriented towards long-term societal objectives and proactively foster a more sustainable economic, social and environmental development (i.e., one that does not lead to economic and financial crisis; that addresses rising social inequalities and respects the natural resources boundaries of the planet). This also includes increasing awareness of and transparency regarding "sustainability" risks that may have an impact on the stability of the financial system.

Regarding the access to external sources of finance, this is often closely related to: (i) the stage of development of the firm (i.e., initial, advance, final stage); (ii) its innovation projects and (iii) the ability of the firm and of the donor to measure the impact and value that a project could create.

The difference in financial access among firms at different development stages depends on their distinct approaches to sustainable entrepreneurship and innovation.

Sustainable entrepreneurship can unfold in established companies (incumbents) as well as in emerging and young companies (start-ups). While well-established firms often improve on radical innovation by investing in incremental innovation processes, radical innovation disproportionately originates in smaller and entrepreneurial new firms (Baumol 2010). Similar findings have also already been established for sustainable innovation (Weiß and Fichter 2013), implying a stronger impact of start-ups in the transition towards a sustainable or green economy.

But what kind of innovation are we talking about?

- *Design innovation*: Thinking with a circular mindset from the initial phases of product design is critical. This can be done through (1) using appropriate materials—recycled and recyclable with a minimal environmental footprint and impact or (2) designing for longevity rather than obsolescence—all the while providing superior performance to the consumer. An example is the start-up firm manufacturing sneakers, Veja[1]

- *Process innovation*: The circular economy or new manufacturing technologies and materials, ranging from 3D printing offering wide-ranging opportunities to significantly reduce waste associated with manufacturing processes. Adidas[2] paired up with Parley for the Oceans[3] to repurpose the millions of pounds of plastic currently polluting the world's oceans. Instead of letting the waste remain, Adidas has found a smart way to use recycling to their (and the planet's) benefit.
- *Materials innovation*: The materials used in the production process are as important as the process itself. Crop-á-porter[4] makes fabric out of crop waste. After farmers harvest pineapples or bananas, the waste is typically burned or left to rot on the ground, emitting greenhouse gases. The new process, which extracts cellulose from the waste to make new fibre and textile, turns the waste into a new source of income for farmers. Clothing made from the fabric can be composted. Adidas prototyped *Primeknit* shoes made from recovered ocean fishing nets while Nike is using recycled polyester, diverting plastic bottles from landfill.
- *Closing loops*: Reusing, ideally upcycling, the materials within product can ensure that the embedded resources find another application again, and again. Examples are the Adidas' technology *Flyleather* material, which is created from cowhide waste mixed in with synthetic fibres. It's led to a decrease in leather being thrown away and 95% of the water used in manufacturing is recycled.
- *Business model innovation*: We are used to buy a new product, what if we no longer pay for buying, but rather pay for access to that product? A few examples of "access over ownership" in the apparel sector are starting to emerge, from Rent the Runway[5] to Mud Jeans.[6] This business model transformation provides new incentives for companies to design longer-lasting products while at the same time providing consumers with the service they are after.

[1] https://www.veja-store.com
[2] https://www.adidas.com/us
[3] https://www.parley.tv/#fortheoceans
[4] https://globalchangeaward.com/winners/crop-a-porter/
[5] https://www.renttherunway.com
[6] https://mudjeans.eu

4.2.1 *Sustainable Start-Ups*

A start-up is a company, a partnership or temporary organisation designed to search for a replicable and scalable business model. Through the start-up phase, new ideas are brought to the market and transformed in economically sustainable enterprises. New firms are artefacts for transforming entrepreneurial choice into profit (Spender 2014). Existing research indicates that forming relationships with external partners is a priority for the success of start-ups (Pangarkar and Wu 2012; Kask and Linton 2013). Sustainable start-ups like any other start-ups are dependent on adequate resource acquisition. Finance is characterised as a central aspect of entrepreneurial success (Schaper 2002).

Sufficient initial capital may provide start-ups with a buffer that enables them to overcome low performance and liquidity difficulties in the early phases (Gimeno et al. 1997). Yet, a start-up may experience difficulty initially when looking for money due to its lack of collateral/revenues, unknown/inexistent credit history and/or radical innovation with no market history or benchmark (Staroßom 2013; Cosh et al. 2009; Kerr and Nanda 2009). Besides these problems, a sustainable start-up might experience further and other challenges due to their involvement in business activities where markets generally do not work well (Patzelt and Shepherd 2011; Di Domenico et al. 2005; York and Venkataraman 2010) and the attempted mobilisation of resources occurring in institutional environments that are not very supportive (Desa 2012). Radical sustainable innovation can take considerable time and effort, which does not necessarily correspond well with expectations of short investment horizons (Randjelovic et al. 2003). The potential conflict between short-term profits and a triple bottom line of economic, environmental and social value creation may create difficulties related to entrepreneur–investor relations and a potential "mission-drift" of the company. How can we solve such a puzzle? Pushing start-ups to provide ESG-proofed business plan/strategy could help them to gather external finance. Using ESG criteria vastly decreases unnecessary risks. The extra filter assesses companies according to their corporate governance, working practices, environmental risks and social impact. If they applied rigorous ESG filters, sustainable start-ups would be better equipped to properly assess their fit in the market. Of course, reports on ESG need to be applied also by the financial institutions when assessing the value of a start-up. The prevalent idea today is that looking at the ESG profile of start-ups would be a deviation from other important indicators, such as founders' profiles, financial models or market benchmarks. But this belief can backfire the whole system of sustainable star-ups, hence it needs to change.

4.2.2 *Existing Firms in Transition to a Sustainable Approach*

The transition of existing firms from a traditional business to adopting a sustainable approach faces different problems than the ones we have seen for sustainable start-ups. Specifically, successful corporate transitions require stewardship, corporate vision and ability to interact with a multiplicity of stakeholders (Keijzers 2002; Loorbach and Wijsman 2013). Some authors highlight that partnering with NGOs is crucial to promote successful radical innovation in the sphere of sustainable development (e.g., Luqmani et al. 2017). From a different angle, the involvement of NGOs and of non-shareholding stakeholders may help avoid greenwashing in event marketing (Griese et al. 2017) where, clearly, the risk of greenwashing is higher for established traditional enterprises than for newly launched start-ups.

In their rather long and extensive study of corporate practices relating to the sustainability discourse, Kiron et al. (2017) identify some major patterns for the transition at established companies. In particular, top management has to a have a vision and ambition towards the transition, develop a strategy to achieve it and, importantly, set up the right organisational structure to accompany the transition.

Overall, it seems that the specification and the structuring of the organisational functions within the enterprise play a major role in the decision of a firm to engage in a sustainable transition strategy and affect the shape and the probability of success of such a strategy.

4.3 Managerial and Organisational Implications for Accessing Green Finance

Sustainable managers, if they really aim to embrace the SDGs challenge, need to change their shareholder-value orientation to one that has a stakeholder-valued perspective.

According to the Ethical Corporation's latest Responsible Business Trends report (2018), 69% of business executives surveyed said they are integrating SDGs in their strategies. At the same time, the number of companies receiving B Corp Certification—which measures a firm's social and environmental performance—has increased in recent years.

According to the report, eight out of ten respondents of the survey stated that sustainability needs to be integrated into the strategies and outputs from Supply Chain to Procurement up to R&D. This should result in increasing sustainable innovations and new product developments (NPDs, henceforth) that align to the Global Goals. Yet, the same report highlights some inconsistency. For example, the same pool of respondents has ranked the most important roles that sustainability covers. The rank shows "reputation preservation and crisis response" (82%) at the top, followed by "Engaging stakeholder groups" (78%), "employee engagement" (72%) and "supply chain management" (72%). Even though all four are crucial to the business and its future success if businesses decide to undergo the required transformation, sustainable managers need to focus more on the last dimension and to offer an increasing role to NPD.

To meet sustainability challenge, a firm's strategy is restrained not only to the interest of stakeholders in the finished product but also to the process to get that final product, that is, supply and value chain.

In other words, to meet the challenge companies need to reach out for responsible production GOAL 12 of the SDG, which in turns requires new managers to implement such change in value and supply chain. The reorientation of value and supply chain is the consequence of the first two features of the innovative firm: strategy and organisation.

Before moving on to how to reach the SDG 12, we would like to share a few thoughts on why it must be done.

Sustainable companies are:

1. Tapping into new and emerging markets: the green revolution is developing new markets and expanding old ones, creating a wealth of new opportunities for ambitious enterprises.
2. Attracting investors: banks, municipalities and even crowdfunding contributors are all eager to support businesses with a sustainable focus.
3. Getting ahead of regulations: staying ahead of environmental regulations helps companies avoid the cost and consequence of a reactive effort if and when those regulations change.
4. Increasing profitability: a serious, sustained and collaborative adoption of sustainable practices has the opportunity to eliminate waste, lowering an organisation's costs.

5. Promoting innovation: operating sustainably requires careful self-scrutiny and in-house auditing that often reveals opportunities to innovate new products and processes.

Of course, to reap those gains the firm is required to show measures, impact and values of its sustainable conduct. Whether it is fair trade and recruitment practices or the impact of climate change and population growth, investors are demanding greater insight into all risk factors.

Company sustainability reports provide this information to some extent, but what is included is largely at the discretion of the business itself. Sometimes the reporting is even promotional in nature and this could represent the risk that companies use the SDGs as a communication tool without much actual adaption to strategy or measurement of their impact towards the Goals.

For this reason, reports, as they are conceived right now, are not an effective way to really measure the sustainability conduct of companies because they, sometimes, do not reveal whether their policies are part of the strategic engagement of the firm or consequences of an internal reorganisation.

Another problem that comes from the company sustainability reports is the fact that many of the sustainability strategies are scattered attempts at introducing spotty measures or activities or procedures within the traditional production process, but without implementing changes of the same process or imagining that the entire process could be sustainable.

The concept of sustainable development needs to be incorporated into the policies and processes of a business if it is to follow sustainable development principles. This does not mean that new management methods need to be invented. Rather, it requires a new cultural orientation and extensive refinements to systems, practices and procedures.

The main areas of the management system that must be changed are those concerned with:

- A greater accountability to non-traditional stakeholders.
- Continuous improvement of reporting practices.
- Moving from Compliance to competitive advantage.

Developing an effective management framework for sustainable development requires addressing both decision-making and governance. The concept of sustainable development must be integrated both with business planning and with management information and control systems.

Governance is increasingly important because of the growing accountability of the corporation and its senior management. Information and reporting systems must support this need. Decision-making at all levels must become more responsive to the issues arising from sustainable development.

Even though commitment of the board's or the CEO's company is an important step towards sustainability, a long-term strategy should be to create an internal position, within the organisational capacity of the company, for sustainable finance manager. A manager that must provide reports that measure performance against these strategies.

Finally, most companies focus on compliance, not competitive advantage—for good reason. In reality, however, they concentrate on ensuring compliance with current environmental regulations, remediating environmental problems caused by past operations, and anticipating the impact of proposed regulations. Sustainable managers should welcome a world in which they could search for win-win solutions because by implementing innovative sustainable practice, they could be ahead of the regulations.

4.4 Preliminary Evidence on the Nexus Between Organisational Structures and ESG Ratings

We repeatedly highlighted the key role played by the shape of the organisational structure of a company in terms of eliciting the company's choice to engage in a sustainable transition strategy, as well as the probability of the strategy being carried out with success. In light of that, here we aim to use some preliminary evidence to ascertain whether the centrality of the sustainability compliance function (CSCF) within a company's organisational chart has implications in terms of the ESG rating that company will obtain.

Our first task was to define a quantifiable approach to measure the centrality of the sustainability compliance function. We defined a metrics—based on information publicly available on companies' websites—hinging on the following two ingredients:

1. Whether the company's organisational chart showed the presence of a sustainability compliance function—in which case the company was assigned value 1, while it received 0 otherwise.
2. Whether the company, besides having or not the sustainability compliance function in its organisational chart, had an internal Committee in charge of sustainability strategy—the company would receive a score of 1 if such Committee existed and it were solely

devoted to the sustainability strategy, it would receive a score of 0.5 if the Committee had another function beside the sustainability strategy, it would receive 0.33 in case it had two more functions, it would receive 0.25 if it had three more functions.

The measure of centrality of the sustainability compliance function (CSCF) was then calculated as the sum of the two ingredients above referring to the 38 listed firms included in the FTSE MIB30 basket for which the appropriate information could be retrieved. The maximum theoretical value was 2, the minimum being 0. While both of the extreme values were actually observed, the mean CSCF resulted equal to 0.71 while the median was at 0.50. The 38 companies were then subdivided into three groups: Group 1 consisting of the 12 companies whose CSCF was 0, Group 2 including the 9 companies whose CSCF was greater than 0 but not greater than 0.50 and Group 3 comprising the remaining 17 companies whose CSCF was greater than 0.50.

Alongside that, we extracted for each one of the 38 companies the ESG rating as freely available on the CSRHUB website (https://www.csrhub.com/). Finally, we calculated the average ESG ratings for each one of the three Groups as defined above.

In Fig. 4.1, we can see that the differences are noticeable. Group 1 includes 12 firms (or 31.6% of the total) and identifies firms where the sustainability compliance function is wholly missing in the organisation chart. Class 2 includes the 9 (or 23.7%) firms in which the sustainability

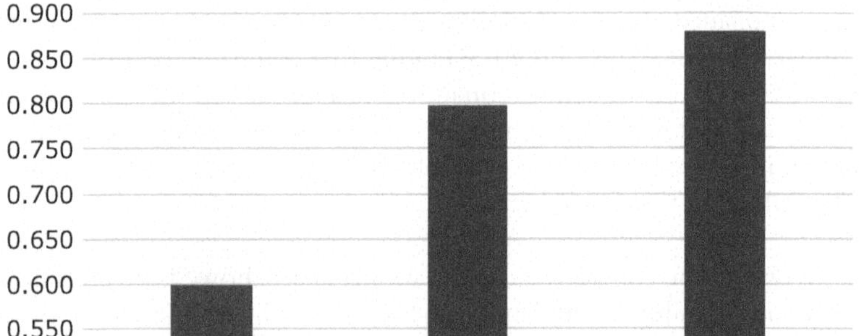

Fig. 4.1 Mean ESG rating by increasing CSCF (Notes: Authors' elaboration)

compliance function reaches an intermediate level in the organisation chart. Finally, Class 3 includes the remaining 17 (or 44.7%) firms where the sustainability compliance function reaches the highest centrality in the organisation chart. It is visible how the average ESG rating keeps increasing if we move from Class 1 (just about 0.6) to Class 2 (close to 0.8) and then to Class 3 (approaching 0.9).

In all, this preliminary evidence suggests that higher values of the CSCF indicator are rewarded in terms of higher ESG ratings. In other words, the companies that are more organisationally compliant to sustainability engagement will indeed have easier access to sustainable investors such as SRI funds and Impact investing institutions.

4.5 Conclusions

In this chapter we have proposed that firms engaging themselves in a credible strategy towards sustainability—either in the initial phase of a start-up or in the transition for an established company—may benefit in terms of access to the expanding pool of finance managed by sustainable investors.

Indeed, referring to some preliminary evidence about the largest listed companies in the Milan Stock Exchange, we have shown that the companies that are more credibly engaged in a sustainable strategy receive systematically higher ESG ratings, where these ratings are the key ingredient to lure in sustainable investors. Therefore, higher ESG-rated companies will have more funding and/or pay lower cost to tap external financial resources.

However, our main point was to argue that in order to suit the Green Finance evolution, a firm has to structure—if a start-up—or reorganise—if already in business—in a way to be sustainability compliant and, what's more, to be able to display that credibly. In turn, this passage demands that a firm adopts a sustainable finance management approach, which may be achieved by hiring sustainable finance managers—especially in medium–large firms—or outsourcing that function to external consulting experts. In other words, we identified in the organisational reshaping of the company—evidently, connected with the role of the sustainable finance manager—the keystone on which the company can credibly show it's engaged in becoming sustainability compliant to external counterparties.

References

Baumol, W. J. (2010). *The microtheory of innovative entrepreneurship*. Princeton/Oxford: Princeton University Press.

Brown, J. R., Fazzari, S. M., & Petersen, B. C. (2009). Financing innovation and growth: Cash flow, external equity, and the 1990s R&D boom. *The Journal of Finance, 64*(1), 151–185.

Cosh, A., Cumming, D., & Hughes, A. (2009). Outside entrepreneurial capital. *The Economic Journal, 119*(540), 1494–1533. https://doi.org/10.1111/j.1468-0297.2009.02270.x.

Desa, G. (2012). Resource mobilization in international social entrepreneurship: Bricolage as a mechanism of institutional transformation. *Entrepreneurship Theory and Practice, 36*(4), 727–751.

Di Domenico, M. L., Haugh, H., & Fichter, P. T. (2005). Social Bricolage: Theorizing social value creation in social enterprises. *Entrepreneurship Theory and Practice, 34*(4), 681–703.

Fichter, K. (2005). *Interpreneurship—Nachhaltigkeitsinnovationen in interaktiven Perspektiven eines vernetzenden Unternehmertums* (Theorie der Unternehmung, Vol. 33). Marburg: Metropolis Verlag.

Gimeno, J., Folta, T., Cooper, A., & Woo, C. (1997). Survival of the fittest? Entrepreneurial human capital and the persistence of underperforming firms. *Administrative Science Quarterly, 42*(4), 750–783. https://doi.org/10.2307/2393656.

Griese, K. M., Werner, K., & Hogg, J. (2017). Avoiding greenwashing in event marketing: An exploration of concepts, literature and methods. *Journal of Management and Sustainability, 7*, 1.

Hall, B. H., & Lerner, J. (2010). The financing of R&D and innovation. *Handbook of the Economics of Innovation, 1*, 609–639.

Hart, O., & Moore, J. (1994). A theory of debt based on the inalienability of human capital. *The Quarterly Journal of Economics, 109*(4), 841–879.

Kask, J., & Linton, G. (2013). Business mating: When start-ups get it right. *Journal of Small Business and Entrepreneurship, 26*(5), 511–536. https://doi.org/10.1080/08276331.2013.876765.

Keijzers, G. (2002). The transition to the sustainable enterprise. *Journal of Cleaner Production, 10*(4), 349–359.

Kerr, W. R., & Nanda, R. (2009). Democratizing entry: Banking deregulations, financing constraints, and entrepreneurship. *Journal of Financial Economics, 94*(1), 124–149.

Kiron, D., Unruh, G., Reeves, M., Kruschwitz, N., Rubel, H., & ZumFelde, A. M. (2017). Corporate sustainability at a crossroads. *MIT Sloan Management Review, 58*(4), 86.

Lazonick, W. (2015). *Stock buybacks: From retain-and-reinvest to downsize-and-distribute*. Washington, DC: Center for Effective Public Management at Brookings.

Loorbach, D., & Wijsman, K. (2013). Business transition management: Exploring a new role for business in sustainability transitions. *Journal of Cleaner Production, 45*, 20–28.

Luqmani, A., Leach, M., & Jesson, D. (2017). Factors behind sustainable business innovation: The case of a global carpet manufacturing company. *Environmental Innovation and Societal Transitions, 24*, 94–105.

Pangarkar and Wu. (2012). Industry globalization and the performance of emerging market firms: Evidence from China. *International Business Review, 21*(2), 196–209.

Patzelt and Shepherd. (2011). The new field of sustainable entrepreneurship: Studying entrepreneurial action linking "what is to be sustained" with "what is to be developed". *Entrepreneurship Theory and Practice, 35*(1), 137–163.

Randjelovic, J., O'Rourke, A. R., & Orsato, R. J. (2003). The emergence of green venture capital. *Business Strategy and the Environment, 12*(4), 240–253.

Schaper, M. (2002). Introduction: The essence of ecopreneurship. *Greener Management International*, (38), 26–30. Retrieved from http://www.jstor.org/stable/greemanainte.38.26

Spender, J. C. (2014). *Business strategy: Managing uncertainty, opportunity, and enterprise*. Oxford: Oxford University Press.

Staroßom, H. (2013). Gründungsunterstützung durch öffentliche Hilfen. In H. Staroßom (Ed.), *Corporate Finance Teil 2* (pp. 73–108). Wiesbaden: Springer.

Weiß, R., & Fichter, K. (2013). *Green Economy Gründungsmonitor*. Konzeptstudie und Piloterhebung. Abschlussbericht. Berlin: Borderstep.

York, J. G., & Venkataraman, S. (2010). The entrepreneur–environment nexus: Uncertainty, innovation, and allocation. *Journal of Business Venturing, 25*(5), 449–463.

CHAPTER 5

Financial Performances of Green Securities

Dirk Schiereck, Gunnar Friede, and Alexander Bassen

5.1 Introduction and Scope

Empirical research in financial markets has been interested in the relationship between the financial performance of an exchange-listed company and its behavior as corporate citizen for a long time. Today, this debate concentrates on the financial performance of sustainable investment projects, and consequently, the following analyses summarize the state of the literature on the performance of the fast-growing segment of green bonds. However, for a deeper understanding of the performance drivers in this segment, we start with a more general overview with respect to the environmental, social, and governance (ESG) performance.

D. Schiereck
Darmstadt University of Technology, Darmstadt, Germany
e-mail: schiereck@bwl.tu-darmstadt.de

G. Friede
University of Hamburg, Hamburg, Germany

DWS Group, Frankfurt am Main, Germany

A. Bassen (✉)
University of Hamburg, Hamburg, Germany
e-mail: Alexander.Bassen@uni-hamburg.de

M. Migliorelli, P. Dessertine (eds.), *The Rise of Green Finance in Europe*, Palgrave Studies in Impact Finance,
https://doi.org/10.1007/978-3-030-22510-0_5

5.2 A Literature Review on the ESG Performances Vis-à-Vis Financial Performances

For many decades, academia and investors have been analyzing environmental, social, and governance (ESG) factors and related concepts like corporate social performance (CSP) in relation to corporate financial performance (CFP). Research on CSP is embedded in the century-long debate about the extent of the social and environmental responsibilities of companies and their actors (Bowen 1953; Carroll et al. 2012; Clark 1916; Freeman 1984; Ghoshal 2005; Lee 2008). In an investor context, the CSP construct is explicitly extended to include environmental and corporate governance elements—resulting in the acronym ESG. Both constructs, CSP and ESG, have at their heart how and in which way shareholder and stakeholder considerations—the business and society relationship—can be balanced (Carroll 1999).

As a field develops, the balance of research work typically shifts from theory generation to theory testing (Kuhn 1962). This pattern also holds in the area of ESG/CSP with an emerging focus on empirical instead of conceptual contributions. Some of the earliest contributions came from Bragdon and Marlin (1972) analyzing if pollution is profitable and from Moskowitz (1972) whether high CSR performers report higher stock returns. A sheer unmanageable number of diverse empirical studies on the ESG-CFP relation have been published since then. An increasing velocity of empirical research is observable since the mid-1990s comparable to the trend toward empirical contributions in economics (Capelle-Blancard and Monjon 2012; Douglas et al. 2008; Lee 2008; Lockett et al. 2006). Based on a relatively small number of primary papers, the findings are often described as ambiguous, inconclusive, or contradictory (Aupperle et al. 1985; Devinney 2009; Griffin and Mahon 1997; Orlitzky 2011; Revelli and Viviani 2015; Rowley and Berman 2000). The reasons being, for instance, the ever-changing definitions and operationalizations of ESG and CFP, a broad range of applied theoretical frameworks, diverse empirical methodologies and data, as well as different levels of quantitative sophistication (Brooks and Oikonomou 2018). The currently prevailing narrative about the relation of ESG and CFP is that the majority of primary studies and meta-analyses reveal a positive association; however, the question of causality, comparable to many social science research, remains less clear. This leaves the supporters advocating for ESG and the skeptics discounting ESG (Peloza 2009). This chapter reflects the state of empirical ESG-CFP research and its main findings supported by two second-order meta-analysis.

Depending on whether the marginal rewards of ESG and CFP deviate from its marginal costs, the ESG-CFP relationship can take different directions and shapes. Diverse theoretical frameworks for the relationship exist. Support for a positive ESG-CFP relation is typically rooted within the management literature in the stakeholder theory (Donaldson and Preston 1995; Freeman 1984), the resource-based theory of the firm (Barney 1991; Barney et al. 2011; Hart 1995; Wernerfelt 1984), or drawn on the literature on competitive advantages (Porter 1979; Porter and Kramer 2006). A negative ESG-CFP relationship is typically based on traditional financial and economic theories, which claim that the only social responsibility of the company is to increase its profits (Friedman 1962, 1970; Levitt 1958). Jensen and Meckling's theory of the firm (1976) reject the notion of social responsibility, as according to their logic a firm does not have responsibilities that only individuals could bear. Arguments based on portfolio theory (Markowitz 1952, 1959) suggest disadvantageous limitations for portfolio construction as a per se exclusion of certain stocks or sectors may lead to suboptimal portfolios. Other preferences than risk/return-considerations, in a narrow sense, are correspondingly unadvisable. Finally, proponents of the efficient market hypothesis argue for a neutral or at least, after costs, non-exploitable relation of ESG and CFP (Fama 1970, 1991; Jensen 1978). According to this hypothesis, all information, including ESG information, is taken into account by investors at all times in the fiction of efficient markets.

Increasingly, indications are found that researchers should not think about the ESG-CFP relation in a linear fashion only (Barnett and Salomon 2006, 2012). It may well be, comparable to other phenomena in economics and management, that the marginal utility of ESG on CFP is decreasing (increasing). The relationship of the two constructs could have a context-specific maximum (minimum), after which a further increase in the cause leads to a decrease (increase) in the outcome. After a certain inflection point the relations could turn asymptotic (exponential) resulting in an overall pattern of curvilinearity. This too-much-of-a-good-thing (TMGT) effect—respectively, the too-little-of-a-good-thing (TLGT) effect—would result in an inverted U-shaped or U-shaped curve of the ESG-CFP relation (Pierce and Aguinis 2013; Trumpp and Guenther 2017). The use of quadratic terms for the proper modeling of non-linear ESG-CFP effects (Gao et al. 2017; Nollet et al. 2016; Trumpp and

Guenther 2017), or of product terms for better consideration of moderating effects (Aguinis et al. 2005), has also increased explanatory power.

As different authors highlight, the ESG-CFP relation is most likely affected by various moderating, mediating, and even confounding factors (Aguinis and Glavas 2012; Peloza 2009). Moderators and mediators fall into categories that influence the inputs of ESG and in particular the outputs of ESG.[1] Various moderating factors of the ESG-CFP relation have been hypothesized as, for example, for firm size (Brammer and Millington 2004; Buehler and Shetty 1974; Johnson and Greening 1999), industry characteristics (Albertini 2013; Griffin and Mahon 1997; Orlitzky et al. 2003), regions (Dixon-Fowler et al. 2013), national institutional contexts (Sandhu et al. forthcoming), financial and economic conditions (Golicic and Smith 2013; Graves and Waddock 1994; Waddock and Graves 1997), stage in the firm life cycle (Elsayed and Paton 2009), research and development investment (McWilliams and Siegel 2000), and visibility with the public or the proximity to consumers (Fry et al. 1982; Jiang and Bansal 2003; Peloza and Shang 2010; Servaes and Tamayo 2013). The more market-based CFP is considered, the list needs to be further extended by the six factors identified in asset pricing theories like systematic market risks, profitability, level of investment, and momentum (Carhart 1997; Fama and French 1993, 2015, 2017; Jegadeesh and Titman 1993). In a second-order meta-analytical context, further potentially moderating factors are testable like study size (Hedges 1981; Trikalinos et al. 2004), study publication year (Borgers et al. 2013; Griffin and Mahon 1997; Rubera and Kirca 2012), publishing outlet impact factor and domain (Orlitzky 2011), or the methodological quality of meta-analyses (Aguinis et al. 2011; Lipsey and Wilson 1993).

Last but not the least, the research into the time lag in which ESG effects have a CFP impact (or vice versa) is an evolving research area. Good ESG performances have been modeled to develop causal impact within days (Flammer 2015), months (Giese and Nagy 2018; Nagy et al. 2016), a year (Lev et al. 2010; McGuire et al. 1988; Waddock and Graves 1997), or several years (Eccles et al. 2014; Edmans 2011; Weber and Gladstone 2014). Investigating the different half-lives of ESG signals in

[1] Mediators are those variables that explain the underlying mechanisms and addresses how, or by what means, an independent variable is able to influence a dependent variable. Whereas moderator variables specify when certain effects will hold and affect the direction and/or strength of the relation between variables (Baron and Kenny 1986).

dependence of the context and the level of ESG data aggregation will further provide interesting research areas. Such longitudinal research context may also offer new insights into potential learning effects of investors on the ESG-CFP relation (Borgers et al. 2013; Derwall et al. 2011). In relations with more sophisticated model specification and better and longer data availability for researchers, the ESG-CFP relation could be further disentangled.

While considerable primary research on ESG-CFP has been conducted, it is less clear how the aggregate picture of these studies look like and how different ESG dimensions and CFP categories interrelate (Aguinis and Glavas 2012; Peloza 2009). Dedicated review studies on the ESG-CFP relation exist, but even the largest previous review study analyzes just a fraction of existing primary studies and thus does not provide a robust overall picture. Summaries of meta-analyses are known in other scientific domains since the beginning of the 1990s (Lipsey and Wilson 1993) to aggregate and reconcile knowledge in a certain field. Since then various other second-order summary formats have emerged. But it is still relatively recent that methods have been developed to allow the more accurate calculation of the second-order average effect and the estimation of its sampling error (Schmidt and Oh 2013).

It has been the focus of two papers to cumulate the fragmented and inconclusive knowledge on the financial effects of ESG criteria in two second-order meta-analyses to derive more robust conclusions and generalizable statements (Busch and Friede 2018; Friede et al. 2015). The 2015 paper blends the results of 35 previous vote-count studies and 25 meta-analyses to the broadest available set of studies and review techniques. It is particularly interested in the set of the following research questions: (1) How many empirical studies on ESG and CFP exist, and what is the relation across vote-count studies and meta-analyses? (2) Do differences of the ESG-CFP relation exist for regions, asset classes, and in the different E, S, and G categories? (3) Do differences of portfolio-based versus non-portfolio-based studies exist? (4) How has the relation of ESG and CFP developed over time? For answering questions 2 and 3, this study in particular focuses on the subsample of primary studies provided by the vote-count studies.

The 2018 paper adds granularity on the CSP-CFP relation and tests various hypotheses about different CSP dimensions and CFP categories. The paper deconstructs the ESG-CFP relation in different CSP dimensions and CFP categories to gain a better understanding on the input-output relation of CSP-CFP and its contextual and statistical robustness. The main research questions are as follows: (1) Which direction and causal relations exhibit the CSP-CFP relationship? (2) Is the relationship between CSP excluding corporate environmental performance (CEP) and CFP stronger than for CEP and CFP? (3) Which relation to CFP exhibits the most frequently analyzed CSP dimensions? (4) Which relation to CSP exhibits the most frequently analyzed CFP categories? (5) Which factors do moderate the CSP-CFP on the second level?

The starting point of the research is the notion in the literature that the ESG-CFP relationship is often described as ambiguous. Both papers argue that the ESG-CFP relation is actually not ambiguous, but to a large degree even significantly positive. The 2015 paper ascertains that more than 90% of studies find a non-negative ESG-CFP correlation, whereas even 47.9% of primary studies in vote-count studies and 62.6% in meta-analyses exhibit a significant positive correlation. Only 6.9% of primary studies in the vote-count sample and 8.0% in the meta-analyses sample find a negative ESG-CFP relation, with the rest finding a non-significant relation. The central-weighted average effect size (correlation) of the ESG-CFP relation, either in vote-count studies ($\bar{\bar{r}}_v$) or meta-analyses ($\hat{\bar{\bar{p}}}_{\mathrm{i}}$), is determined at 0.146 and 0.150, respectively. The average effect size reported for different primary studies remained on average unchanged since the 1990s, suggesting limited learning effects of investors on the ESG-CFP relation. Looking in more detail on the results of the vote-count studies, the positive ESG-CFP relation holds across asset classes, regions, and the individual E, S, and G categories. A disproportionate positive relation is detected for studies focused on North America, the emerging markets, and non-equity asset classes like bonds and real estate. Moreover, a considerable difference between portfolio-based and non-portfolio-based studies is discovered, which may contribute to a potentially distorted perception among investors on the ESG-CFP relation.

The 2018 paper examines the sample of the 25 first-order meta-analyses in greater detail. It finds a highly significant, positive, and bilateral CSP-CFP relation, providing support for the virtuous circle hypothesis. The findings are robust at first-order and second-order level. Various checks for potential publication bias and sensitivity analyses do not alter the findings significantly. The positive CSP-CFP correlation holds for various CSP dimensions, in particular for corporate reputation and philanthropy, but also for various CFP dimensions like perceptional, operational, and accounting-based performance. Finally, the data is subject to various robustness and moderator checks. The paper cannot confirm that social issues in management (SIM) journals or methodological weaker papers tend to publish higher CSP-CFP effect sizes, nor that publication bias distorts the results. Moreover, the lower effect sizes for more recent meta-analyses are explained by higher meta-analytical attenuation factors and are not mirrored by a shrinkage of effect sizes on primary study level.

The analyses make several contributions for research and practice. Considering the intense debate of scholars from different schools of thought during the last decades, the evidence for at least a non-negative ESG-CFP relation is striking, challenging the traditional trade-off logic. Shareholder theories' mantra of a win-lose relation of ESG and CFP (Friedman 1962, 1970; Levitt 1958) fails empirical evidence. The results of the two second-order meta-analyses therefore falsify, at the point of time, claims of a negative association and causation of ESG and CFP. This has potential implications for the objective function of managers, corporates, and investors. The ex-ante assumption that agency costs of executives (Berle and Means 1932; Jensen and Meckling 1976) could be reduced through the sole objective to maximize shareholder return (Jensen 2001; Sundaram and Inkpen 2004) is based on a weak empirical foundation. This understanding enables an enlightened corporate objective for optimizing collective value—for shareholders and stakeholders alike (Donaldson and Walsh 2015). As more and more investors realize, purpose and profits are inextricably linked (Fink 2019).

Embracing stakeholder theories' idea of creating win-win relations among all stakeholders, including investors (Freeman et al. 2007), could significantly reorient the opportunity set of companies, investors, and increase planetary welfare overall. Decades of empirical research should provide some confidence that the cooperative (stakeholder-oriented) approach of the classical prisoners' dilemma (Granovetter 1985; Jones 1995) can also be transferred into the area of stakeholder-oriented management and investment.[2] Realizing that stakeholder wealth can be posi-

[2] Cooperative behavior, even under individual competition, was shown to maximize the collective welfare in the classical prisoners' game (Axelrod and Hamilton 1981; Nowak and Sigmund 1993). This is more valid for repeated games, making cooperative behavior the most reasonable option for long-term human interaction (Hofbauer and Sigmund 1998). As Nowak (2006) points out, cooperation therefore is and has been the default option not only in cells, social insects, but also in human society. The empirical results suggest this to be potentially valid also for stakeholder-oriented investments.

tively affected by good ESG performance at no measurable shareholder disadvantage (CFP) may increase investor support for ESG integration and could prove as a catalyst for moving existing ESG integration barriers.

However, stakeholder-oriented management cannot mask bad corporate strategy and execution, nor it can be overburdened rescuing already financially troubled companies. It can also not act as a management excuse for failing short- and long-run financial targets. Even if created for very different purposes, the long-term reason for the existence of private companies is to earn profits above its cost of capital. As Carroll summarizes, fulfilling the firm's economic responsibilities is the foundation upon which all other CSR activities can rest (1991). Necessary profits and fulfilling a corporate purpose are however inextricably linked (Fink 2019).

5.3 The Green Bond Premium

As mentioned in the last section, investing in sustainable assets is often related to a superior performance and a lower risk, ***including studies of sovereign and corporate bonds.*** Consequently, a number of studies transferred ESG findings to ***a subset of bond studies***—the expectations for the performance of green bonds. However, this transfer is challenged by some severe methodological issues. To correctly price a bond and to benchmark the performance of bonds is overall critical because each bond is very specific with respect to credit rating, coupon, maturity, covenants, and further conditions (Maul and Schiereck 2018). Bachelet et al. (2019) put this dilemma precisely in a nutshell: "the best methodological approach for testing this research question is unusable since the counterfactual information (what would have happened if the same bond would not be 'green'?) is not available. The second best to a full-fledged randomized experiment is also impossible since it requires that both treatment and control, respectively green and brown bonds, are issued after the experiment starts and with ad-hoc procedures that satisfy randomness". Given these limitations, all empirical studies use matching procedures and econometric instruments to estimate green bond premiums. While the green bond market has grown significantly during the last couple of years, it still represents a niche market (Sartzetakis 2019) and each analysis has to address issues like illiquidity and price quality when it applies a matching procedure (Bao et al. 2011; Wulandari et al. 2018; Leng and Noronha 2019).

The future success of becoming an important contributor to financial markets and sustainable investments will depend—among others—on pricing and performance of green bonds and therefore on matching procedures and correct econometric implementations. Current support for appropriate pricing is provided by the benchmark bond that now exists in the euro area. In May 2019, the Netherlands became the latest European country to arouse investors' interest in green debentures, preceded by Poland in 2016, followed by France, Belgium, and Ireland. The Dutch finance ministry issued the first green bond at all of a triple A-rated European government. The government announced to use the proceeds raised via the bond to finance sustainable projects in the years ahead. The bond will mature in 2040 and was placed in the market with a spread of 18 basis points above regular Dutch government bonds with similar maturity. This means that investors generate an additional premium for investing in a nearly risk-free sustainable asset. Denmark and Germany announced to be the next European governmental issuers of green bonds. In both cases, market participants also expect to observe yields slightly above regular issues. However, this additional primary market return is not always observed in other bond market segments.

Pricing of green bonds versus non-green bonds beyond triple A-rated government bonds has so far been addressed predominantly in research from investment banks, advisory firms, and the like. In these early analyses, a few bonds are compared to decide if bonds trade "cheap" or "rich". Trading strategies are outlined (Ridley et al. 2016) or indices compared (Preclaw and Bakshi 2015), but the entire population of green bonds has hardly been examined so far and existing studies vary in design and results (Bloomberg 2017). The following survey will clarify this inconclusive picture by summarizing the most recent studies in chronological order and increasing sample sizes, comparing green-labeled and non-green-labeled bonds of the same issuers, thereby complementing the literature examining the pricing of ESG instruments compared to conventional assets.

Hachenberg and Schiereck (2018) examine secondary market prices and compare green bonds with conventional bonds to evaluate the attractiveness of their risk-return profiles. They use data from Bloomberg, in particular the entire population of August 2016 outstanding green-labeled bonds, which delivers an initial data sample size of 732. Taking into account factors such as liquidity and the availability of comparable conventional bonds, they only have a relatively small final sample size of 63 green bonds. Instead of using bond yields they consider Bloomberg's i-spreads. I-spreads are noted in basis points above a risk-free benchmark, usually the swap rate. I-spreads separate the interest and credit part of the spread. By using them in their analysis the authors only look at the differences between green and conventional bonds in terms of the credit part of the yield. They match the daily i-spreads of green-labeled bonds with the i-spreads of similar non-green-labeled bonds using bond triplets, that is, two conventional bonds with the closest maturities are assigned to each green bond. Other than that, the comparable bonds have to exhibit the same issuer, ranking, currency, and structure. For each green bond a synthetic conventional bond of the same maturity is calculated by using linear interpolation of the two matched conventional bonds. The authors then use panel regressions on the i-spread differentials between the green bonds and the comparable, synthetic non-green bonds.

Hachenberg and Schiereck (2018) find that green bonds trade in secondary markets with a small negative i-spread premium overall. After grouping the green bonds into rating categories of AAA-BBB, it turns out that AAA-rated green bonds are traded at a positive premium compared to their corresponding non-green bonds (as in the case of the abovementioned Dutch green government bond), while all non-AAA-rated bonds are traded at a negative premium. The negative premium of AA-, A- and BBB-rated green bonds compared to conventional bonds could compensate for an issuer's external costs. Additionally, the premium is particularly negative for financial and corporate green bonds. They find that neither issue size, maturity, nor currency, but rather government-related and financial issuers and the existence of an ESG issuer rating, has a significant influence on pricing differences.

Kuhn et al. (2018) consider the primary market, in particular all bond issues in the green bond segment of the London Stock Exchange or the Luxembourg Stock Exchange until the end of September 2017. For the composition of the bond data, they use the financial database of Thomson Reuters and the securities prospectus. Untypical emissions and bonds without a credit spread over a risk-free bond are subtracted from the initial data sample size, which leaves plain vanilla fixed-rated bonds only. The final data sample consists of 81 green bonds. Kuhn et al. (2018) apply classic linear regression models to explain a green bond's initial spread. The results indicate that the yield in the primary market can largely be explained by credit risk, the existence of a call option, and the currency. Furthermore, the emission spread of green bonds seems to lower in times of economic growth.

Bachelet et al. (2019) analyze a sample of 89 green bonds spanning the period of 2013 until 2017 and take a focus on secondary markets. Bonds are defined as green when they are listed on the Climate Bonds Initiative website. The authors use a matching procedure based on the characteristics of issued amount, including +/− 400%, coupon rate +/− 0.25%, maturity date +/− 2 years, same currency, same issuer, same rating, as well as same coupon type. They look at the difference in yields of the matched bonds in a bond-day-level panel regression model. Their main finding is that green bonds overall exhibit surprisingly higher yields compared to their conventional counterparts, as well as lower variances and higher liquidity. Considering the issuer and verification process, the authors discover that institutional green bonds (bonds issued by national government, municipality, or supranational institution, such as the World Bank) display negative premiums, whereas private green bonds show positive premiums. Green bonds from private issuers have positive premiums with respect to their brown correspondents, unless the issuer certifies the bond's "greenness". As a result, especially non-certified bonds exhibit particularly positive premiums. The general implication of this evidence highlights the importance of the issuer's reputation or third-party verifications to avoid the suspicion of green-washing and provide favorable financing conditions for the issuing company.

Kapraun and Scheins (2019) investigate the green bond premium over conventional bonds using roughly 1500 green bonds issued between 2009 and 2018. Bonds are included into the dataset according to the Climate Bonds Initiative website, as well as bonds from Reuters are classified as green and bonds from Bloomberg whose "Use of proceeds" contains the word "green". The authors look both at primary and secondary markets. For the primary market, they estimate regressions on the yield at issuance of more than 1500 green bonds and 200,000 conventional bonds. They find a negative green bond premium of 20–30 bps. This premium varies across currencies and issuer type. As the major determinants of the premium size the authors name issuer credibility. Bonds issued by more credible institutions, like governments or supranationals, as well as bonds backed by a collateral are issued at lower yields compared to corporate issuers. In their secondary market analysis, the authors look at 4617 green-conventional bond couples of the same issuer, with the same rating, seniority, currency, and bond type. The main reason for their large dataset compared to other studies is that they use bond pairs only instead of triplets and therefore only need bid-ask midpoints of one comparable bond. Other studies use bond triplets instead to eliminate any maturity bias. Their key finding is a stable yield difference of around −10 bps between green and conventional bonds, as well as significantly lower yields for bonds traded on dedicated green bond segments, underlining the importance of these segments as market catalysts.

Zerbib (2019) estimates the yield differential between green and conventional bonds using a matching method for 110 green bonds for the time period 2013 to 2017. The green bond sample represents the final sample of the entire population of 1065 green bonds complying with the Green Bond Principles indexed by Bloomberg on December 31, 2017. For the matching procedure, the author uses bond triplets, that is, two conventional bonds matched with one green bond. The two conventional bonds are selected if they are from the same issuer, currency, rating, bond structure, seniority, collateral, and coupon type. Maturity has to fall within +/− 2 years, as well as further matching requirements regarding the bond liquidity have to be fulfilled. In the second stage, the maturity bias is eliminated by interpolating between the two conventional bonds linearly to obtain a synthetic conventional bond yield. Using this synthetic bond yield, the yield difference to the matching green bond can be calculated. Due to the comprehensive matching procedure, the author is left with 110 bonds out of the original 1065 Green Bond Principles-indexed green bonds by Bloomberg, representing 10% or 17% of total outstanding green bonds or total outstanding debt. The green bond premium is estimated as the unobservable effect in a panel regression on the return difference between green and synthetic conventional bonds, controlling for the remaining liquidity difference of both bonds. This estimation provides a small, but significant green bond premium of −1.76 bps. As a result of a subsample analysis, the evidence indicates that the green bond premium is only significant in the financial sector, for EUR and USD denominated, as well as AA-rated bonds. In order to investigate the determinants of the green bond premium, multivariate regression analyses are applied on the premium. The only significant determinants are rating categories AA or A, with any other potential determinant, such as currency, maturity or sector affiliation, being insignificant.

The main summarizing result of the few published studies on green bond premiums is typical for a young, still emerging financial instrument and debt market segment. Based on a very limited number of observations, the findings are inconclusive (probably) depending on the methods applied and the selection criteria of the data sets. The sign of the potential green premium, the so-called *greenium*, has not even been clarified (Kapraun and Scheins 2019). However, the premiums seem to correlate with the credit risk of an issue expressed by its rating and a third-party verification. From an investor's point of view, one next step in empirical research has to focus on recalculating green bond premiums to stabilize the knowledge base about the size and direction of the premiums and to detect trends over time.

From an issuer's perspective, there is another important and still neglected aspect that provides a path for future research on the overall consequences of green bond issuances as changes in the cost of debt. The identification of a green bond premium can be interpreted as a change in the financing cost for one part of the overall investment projects of an issuer. The question remains whether the following brown bonds of a green bond issuer will yield at higher rates. Bachelet et al. (2019) underline the importance of third-party verification. And Tolliver et al. (2019) provide evidence that many post-issuance reports do not clearly identify the additionality of green bond impacts. If investors believe that there is no pure additional green impact, they might raise expectations about partly financing substitutions by green bonds. In the past, investors in corporate bonds financed capital expenditures that were partly green. With the separate issuance of green bonds, investors might perceive that this mix is split up resulting in a pure green and pure brown investment. Theoretical considerations based on value additivity suggest that the combined effect should be captured to address the complete effects of green bond issues. In the case of a financing substitution through green bond issuances, the yields of brown bonds could increase to reflect the now higher carbon risk of the remaining investment projects (Oestreich and Tsiakas 2015). For an overall evaluation of green bond issues on the cost of debt, empirical studies have to compare financing conditions over time and differentiate the overall cost of debt before and after the first issues of green bonds.

For both interested parties, investors and issuers, the challenge also has to be addressed how the specific green bond performance is related to the more general ESG performance. At a first impression, there seems to be a positive correlation. Recent studies about the corporate bond market underline that bonds with high composite ESG ratings have tighter spreads and tend to outperform their peers with lower ESG ratings (Polbennikov et al. 2016). Likewise lenders demand significantly higher interest rates for loans of companies with environmental concerns (Chava 2014). But research in this area also shows that findings are not always positive. There is evidence as well that socially responsible firms do not have lower cost of public debt (Menz 2010). To contribute to this discussion, further analyses on the pricing of green bonds in comparison to conventional bonds have to focus on ESG ratings as explanatory factor.

5.4 Primary Market or Secondary Market: Where Is the Benefit for the Issuer of a Green Bond?

Issuers realize their yields in the primary bond market, but the secondary bond market is nevertheless at least as important as the primary segment. While primary market yields express a market price at one point of time, which can be influenced by randomly distributed imbalance of demand and supply, secondary markets signal the stability of premiums and also indicate windows of opportunity to issue new green bonds.

However, knowledge on the interaction of primary and secondary green bond markets is still sparse and the quality of secondary market prices depends on the liquidity of this emerging segment. Kapraun and Scheins (2019) claim to be the first and only who parallelly examine primary and secondary market effects. Their result that green bonds listed on secondary markets with a dedicated green bond segment are traded on average 20 basis points lower indicates the importance of transparent secondary markets for good primary market conditions. This finding underlines that issuers benefit from standard setting and the reduction of information asymmetry on the secondary market level as this will obviously influence primary market yields.

5.5 Unobserved Reputational Gain

Besides the correlations between the long-run sustainability performance and the financial performance on individual company level, the issuance of green bonds might result in some indirect positive effects. To understand potential reputational effects from the issuance of green bonds, it is necessary to proxy the difficulties to fulfill reporting requirements in the aftermath of the issuance and to estimate capital market reactions at the announcement of green news.

The issuance of green bonds requires some reporting duties with respect to the use of the issuance proceeds. Eccles and Krzus (2019) show for 15 of the largest US oil and gas companies that they are well able to fulfill the recommendations of the Task Force for Climate-related Financial

Disclosures (TCFD). First, this observation is interpreted as a convincing indicator that it is feasible to follow the TCFD's recommendations if companies are interested in doing so. Second, if it is feasible to follow the recommendations, it seems to be also feasible to follow the reporting requirements because of green bond issues. Additionally, Tolliver et al. (2019) examine the extent to which green bond proceeds were allocated to projects and assets aligned with Sustainable Development Goals (SDG) and Nationally Determined Contributions (NDC)-related environmental outcomes. While international finance institutions usually already provide transparent post-issuance assessments, other issuers still leave some room for improvement (see also Schneeweiß 2019). This often unclear use of the proceeds (also with respect to the additionality of green bond impacts) can limit the reputational effects for corporate issuers and can result in some problems for later empirical examinations of indirect green bond effects.

Notwithstanding the still existing shortcomings in post-issuance reporting, the potential for indirect reputational effects obviously is given but hard to quantify. Oestreich and Tsiakas (2015) provide evidence in favor of a carbon risk factor in the stock returns of German companies. Companies with high reported carbon emissions are confronted with a higher exposure to carbon risk and have to offer higher stock returns to compensate for this risk factor. Consequently, a more sustainable firm policy seems to be also an instrument to absorb macroeconomic shocks like changes in the regulatory environment and negative reputational spillovers in cases of news on environmental damages. Consistently, Ramiah et al. (2013) document the impact of 19 announcements of environmental regulation on the systematic risk implied in stock returns listed on the Australian Stock Exchange over a period from 2005 to 2011.

In line with these considerations, Hamilton (1995) and Klassen and McLaughlin (1996) show that negative news in the form of reported high-level toxic emissions result in negative stock returns, which are the larger the poorer the environmental management practices of a company are. The issuance of a green bond and the transparent post-issuance reporting can be interpreted as credible signals of strong environmental management practices.

5.6 Concluding Remarks

While the positive relationship between financial and ESG performance seems to be common sense in empirical research, it is by far too early to state a similar correlation for green bond premiums. Some studies report a positive financial effect by the issuance of green bonds on the overall cost of debt, but this evidence is weak at best. Additionally, the effect seems to be not monotonic over rating classes with triple A-rated government bonds generating higher yields than their conventional counterparts do. This inconclusive evidence indicates the beginning of a long-lasting empirical research agenda as documented for the analysis of financial performance related to ESG performance.

References

Aguinis, H., & Glavas, A. (2012). What we know and don't know about corporate social responsibility: A review and research agenda. *Journal of Management, 38*(4), 932–968.

Aguinis, H., Beaty, J. C., Boik, R. J., & Pierce, C. A. (2005). Effect size and power in assessing moderating effects of categorical variables using multiple regression: A 30-year review. *Journal of Applied Psychology, 90*(1), 94–107.

Aguinis, H., Dalton, D. R., Bosco, F. A., Pierce, C. A., & Dalton, C. M. (2011). Meta-analytic choices and judgment calls: Implications for theory building and testing, obtained effect sizes, and scholarly impact. *Journal of Management, 37*(1), 5–38.

Albertini, E. (2013). Does environmental management improve financial performance? A meta-analytical review. *Organization & Environment, 26*(4), 431–457.

Aupperle, K. E., Carroll, A. B., & Hatfield, J. D. (1985). An empirical examination of the relationship between corporate social responsibility and profitability. *Academy of Management Journal, 28*(2), 446–463.

Axelrod, R., & Hamilton, W. (1981). The evolution of cooperation. *Science, 211*(4489), 1390–1396.

Bachelet, M. J., Becchetti, L., & Manfredonia, S. (2019). The Green Bonds Premium Puzzle: The Role of Issuer Characteristics and Third-Party Verification. Sustainability, *11*(4), 1098.

Bao, J., Pan, J., & Wang, J. (2011). The illiquidity of corporate bonds. *The Journal of Finance, 56*(3), 911–945.

Barnett, M., & Salomon, R. (2006). Beyond dichotomy: The curvilinear relationship between social responsibility and financial performance. *Strategic Management Journal, 27*(11), 1101–1122.

Barnett, M. L., & Salomon, R. M. (2012). Does it pay to be really good? Addressing the shape of the relationship between social and financial performance. *Strategic Management Journal, 33*(11), 1304–1320.

Barney, J. B. (1991). Firm resources and sustained competitive advantage. *Journal of Management, 17*(1), 99–120.

Barney, J. B., Ketchen, D. J., & Wright, M. (2011). The future of resource-based theory: Revitalization or decline? *Journal of Management, 37*(5), 1299–1315.

Baron, R. M., & Kenny, D. A. (1986). The moderator-mediator variable distinction in social psychological research: Conceptual, strategic, and statistical considerations. *Journal of Personality and Social Psychology, 51*(6), 1173–1182.

Berle, A. A., & Means, G. C. (1932). *The modern corporation and private property* (1968th, 10th ed.). New Brunswick: Transaction Publishers.

Bloomberg New Energy Finance Report. (2017). *Investors are willing to pay a 'green' premium*. New York: Bloomberg Finance LP.

Borgers, A., Derwall, J., Koedijk, K., & ter Horst, J. (2013). Stakeholder relations and stock returns: On errors in investors' expectations and learning. *Journal of Empirical Finance, 22*(June), 159–175.

Bowen, H. R. (1953). *Social responsibilities of the businessman*. Iowa City: University of Iowa Press, (Reprint 2013).

Bragdon, J. H., & Marlin, J. A. T. (1972). Is pollution profitable? *Risk Management, 19*, 9–18.

Brammer, S., & Millington, A. (2004). The development of corporate charitable contributions in the UK: A stakeholder analysis. *Journal of Management Studies, 41*(8), 1411–1434.

Brooks, C., & Oikonomou, I. (2018). The effects of environmental, social and governance disclosures and performance on firm value: A review of the literature in accounting and finance. *The British Accounting Review, 50*(1), 1–15.

Buehler, V. M., & Shetty, Y. K. (1974). Motivations for corporate social action. *Academy of Management Journal, 17*(4), 767–771.

Busch, T., & Friede, G. (2018). The robustness of the corporate social and financial performance relation: A second-order meta-analysis. *Corporate Social Responsibility and Environmental Management, 25*(4), 583–608.

Capelle-Blancard, G., & Monjon, S. (2012). Trends in the literature on socially responsible investment: Looking for the keys under the lamppost. *Business Ethics: A European Review, 21*(3), 239–250.

Carhart, M. M. (1997). On persistence in mutual fund performance. *Journal of Finance, 52*(1), 57–82.

Carroll, A. B. (1991). The pyramid of corporate social responsibility: Toward the moral management of organizational stakeholders. *Business Horizons, 34*(4), 39–48.

Carroll, A. B. (1999). Corporate social responsibility: Evolution of a definitional construct. *Business & Society, 38*(3), 268–295.

Carroll, A. B., Lipartito, K. J., Post, J. E., & Werhane, P. H. (2012). *Corporate responsibility: The American experience* (K. E. Goodpaster, Ed.). New York: Cambridge University Press.

Chava, S. (2014). Environmental externalities and cost of capital. *Management Science, 60*, 2223–2247.

Clark, J. M. (1916). The changing basis of economic responsibility. *Journal of Political Economy, 24*(3), 209–229.

Derwall, J., Koedijk, K., & Ter Horst, J. (2011). A tale of values-driven and profit-seeking social investors. *Journal of Banking & Finance, 35*(8), 2137–2147.

Devinney, T. M. (2009). Is the socially responsible corporation a myth? The good, the bad, and the ugly of corporate social responsibility. *Academy of Management Perspectives, 23*(2), 44–56.

Dixon-Fowler, H. R., Slater, D. J., Johnson, J. L., Ellstrand, A. E., & Romi, A. M. (2013). Beyond "does it pay to be green?" a meta-analysis of moderators of the CEP–CFP relationship. *Journal of Business Ethics, 112*(2), 353–366.

Donaldson, T., & Preston, L. E. (1995). The stakeholder theory of the corporation: Concepts, evidence, and implications. *The Academy of Management Review, 20*(1), 65.

Donaldson, T., & Walsh, J. P. (2015). Toward a theory of business. *Research in Organizational Behavior, 35*, 181–207.

Douglas, A. C., Millis, J. E., Niang, M., Stepchenkova, S., Byun, S., Ruffini, C., Lee, S. K., Loutfi, J., Lee, J.-K., Atallah, M., & Blanton, M. (2008). Internet addiction: Meta-synthesis of qualitative research for the decade 1996–2006. *Computers in Human Behavior, 24*(6), 3027–3044.

Eccles, R., & Krzus, M. (2019). Implementing the task force on climate-related financial disclosure recommendations: An assessment of corporate readiness. *Schmalenbach Business Review, 71*, 287–293.

Eccles, R. G., Ioannou, I., & Serafeim, G. (2014). The impact of corporate sustainability on organizational processes and performance. *Management Science, 60*(11), 2835–2857.

Edmans, A. (2011). Does the stock market fully value intangibles? Employee satisfaction and equity prices. *Journal of Financial Economics, 101*(3), 621–640.

Elsayed, K., & Paton, D. (2009). The impact of financial performance on environmental policy: Does firm life cycle matter? *Business Strategy and the Environment, 18*(6), 397–413.

Fama, E. F. (1970). Efficient capital markets: A review of theory and empirical work. *The Journal of Finance, 25*(2), 383–417.

Fama, E. F. (1991). Efficient capital markets: II. *The Journal of Finance, 46*(5), 1575–1617.

Fama, E. F., & French, K. R. (1993). Common risk factors in the returns on stocks and bonds. *Journal of Financial Economics, 33*(1), 3–56.

Fama, E. F., & French, K. R. (2015). A five-factor asset pricing model. *Journal of Financial Economics, 116*(1), 1–22.

Fama, E. F., & French, K. R. (2017). International tests of a five-factor asset pricing model. *Journal of Financial Economics, 123*(3), 441–463.

Fink, L. D. (2019). *Larry Fink's 2019 letter to CEOs. Purpose & Profit*. Available at https://www.blackrock.com/corporate/investor-relations/larry-fink-ceo-letter. Accessed 17 Jan 2019.

Flammer, C. (2015). Does corporate social responsibility Lead to superior financial performance? A regression discontinuity approach. *Management Science, 61*(11), 2549–2568.

Freeman, R. E. (1984). *Strategic management: A stakeholder approach.* Boston: Pitman.

Freeman, R. E., Harrison, J. S., & Wicks, A. C. (2007). *Managing for stakeholders: Survival, reputation, and success.* New Haven/London: Yale University Press.

Friede, G., Busch, T., & Bassen, A. (2015). ESG and financial performance: Aggregated evidence from more than 2000 empirical studies. *Journal of Sustainable Finance & Investment, 5*(4), 210–233.

Friedman, M. (1962). *"Capitalism and freedom", reprint.* Chicago/London: The University of Chicago Press.

Friedman, M. (1970, September 13). The social responsibility of business is to increase its profits. *The New York Times Magazine*, 32–33.

Fry, L. W., Keim, G. D., & Meiners, R. E. (1982). Corporate contributions: Altruistic or for-profit? *Academy of Management Journal, 25*(1), 94–106.

Gao, Y., Wu, J., & Hafsi, T. (2017). The inverted U-shaped relationship between corporate philanthropy and spending on Research and Development: A case of complementarity and competition moderated by firm size and visibility. *Corporate Social Responsibility and Environmental Management, 24*(6), 465–477.

Ghoshal, S. (2005). Bad management theories are destroying good management practices. *Academy of Management Learning & Education, 4*(1), 75–91.

Giese, G., & Nagy, Z. (2018). *How markets price ESG. Have changes in ESG scores affected stock prices?*. MSCI ESG Research LLC.

Golicic, S. L., & Smith, C. D. (2013). A meta-analysis of environmentally sustainable supply chain management practices and firm performance. *Journal of Supply Chain Management, 49*(2), 78–95.

Granovetter, M. (1985). Economic action and social structure: The problem of embeddedness. *American Journal of Sociology, 91*(3), 481–510.

Graves, S., & Waddock, S. (1994). Institutional owners and corporate social performance. *Academy of Management Journal, 37*(4), 1034–1046.

Griffin, J. J., & Mahon, J. F. (1997). The corporate social performance and corporate financial performance debate: Twenty-five years of incomparable research. *Business & Society, 36*(1), 5–31.

Hachenberg, B., & Schiereck, D. (2018). Are green bonds priced differently from conventional bonds?. *Journal of Asset Management, 19*(6), 371–383.

Hamilton, J. T. (1995). Pollution as news: Media and stock market reactions to the toxics release inventory data. *Journal of Environmental Economics and Management, 28*(1), 98–113.

Hart, S. L. (1995). A natural-resource-based view of the firm. *Academy of Management Review, 20*(4), 986–1014.

Hedges, L. V. (1981). Distribution theory for Glass's estimator of effect size and related estimators. *Journal of Educational and Behavioral Statistics, 6*(2), 107–128.

Hofbauer, J., & Sigmund, K. (1998). *Evolutionary games and population dynamics*. Cambridge: Cambridge University Press.

Jegadeesh, N., & Titman, S. (1993). Returns to buying winners and selling losers: Implications for stock market efficiency. *The Journal of Finance, 48*(1), 65–91.

Jensen, M. C. (1978). Some anomalous evidence regarding market efficiency. *Journal of Financial Economics, 6*(2/3), 95–101.

Jensen, M. C. (2001). Value maximization, stakeholder theory, and the corporate objective function. *Journal of Applied Corporate Finance, 14*(3), 8–21.

Jensen, M. C., & Meckling, W. H. (1976). Theory of the firm: Managerial behavior, agency costs and ownership structure. *Journal of Financial Economics, 3*(4), 305–360.

Jiang, R. J., & Bansal, P. (2003). Seeing the need for ISO 14001. *Journal of Management Studies, 40*(4), 1047–1067.

Johnson, R., & Greening, D. (1999). The effects of corporate governance and institutional ownership types on corporate social performance. *Academy of Management Journal, 42*(5), 564–576.

Jones, T. M. (1995). Instrumental stakeholder theory: A synthesis of ethics and economics. *Academy of Management Review, 20*(2), 404–437.

Kapraun, J., & Scheins, C. (2019). *(In)-credibly green: Which bonds trade at a green bond premium?*. Available at SSRN: https://ssrn.com/abstract=3347337

Klassen, R. D., & McLaughlin, C. P. (1996). The impact of environmental management on firm performance. *Management Science, 42*(8), 1199–1214.

Kuhn, T. S. (1962). *The structure of scientific revolutions* (2nd ed.). Chicago: University of Chicago Press.

Kuhn, D., Kiesel, F., & Schiereck, D. (2018). Determinanten von Credit Spreads in Green Bonds in europäischen Emissionsmärkten (No.110452). Darmstadt Technical University, Department of Business Administration, Economics and Law, Institute for Business Studies (BWL).

Lee, M.-D. (2008). A review of the theories of corporate social responsibility: Its evolutionary path and the road ahead. *International Journal of Management Reviews, 10*(1), 53–73.

Leng, F., & Noronha, G. (2019). Relative value in corporate bond sectors. *Review of Quantitative Finance and Accounting, 52*, 717–735.

Lev, B., Petrovits, C., & Radhakrishnan, S. (2010). Is doing good good for you? How corporate charitable contributions enhance revenue growth. *Strategic Management Journal, 31*(2), 182–200.

Levitt, T. (1958). The dangers of social responsibility. *Harvard Business Review, 36*(5), 41–50.

Lipsey, M. W., & Wilson, D. B. (1993). The efficacy of psychological, educational, and behavioral treatment: Confirmation from meta-analysis. *American Psychologist, 48*(12), 1181–1209.

Lockett, A., Moon, J., & Visser, W. (2006). Corporate social responsibility in management research: Focus, nature, salience and sources of influence∗. *Journal of Management Studies, 43*(1), 115–136.

Markowitz, H. (1952). Portfolio selection. *Journal of Finance, 7*(1), 77–91.

Markowitz, H. (1959). *Portfolio selection. Efficient diversification of investments* (1970th ed.). New Haven/London: Yale University Press.

Maul, D., & Schiereck, D. (2018). The market timing of corporate bond reopenings. *The European Journal of Finance, 24*(9), 714–734.

McGuire, J. B., Sundgren, A., & Schneeweis, T. (1988). Corporate social responsibility and firm financial performance. *Academy of Management Journal, 31*(4), 854–872.

McWilliams, A., & Siegel, D. (2000). Corporate social responsibility and financial performance: Correlation misspecification? *Strategic Management Journal, 21*(5), 603–609.

Menz, K.-M. (2010). Corporate social responsibility: Is it rewarded by the corporate bond market? A critical note. *Journal of Business Ethics, 96*, 117–134.

Moskowitz, M. (1972). Choosing socially responsible stocks. *Business and Society Review, 1*(1), 71–75.

Nagy, Z., Kassam, A., & Lee, L.-E. (2016). Can ESG add alpha? An analysis of ESG tilt and momentum strategies. *The Journal of Investing, 25*(2), 113–124.

Nollet, J., Filis, G., & Mitrokostas, E. (2016). Corporate social responsibility and financial performance: A non-linear and disaggregated approach. *Economic Modelling, 52*, 400–407.

Nowak, M. A. (2006). Five rules for the evolution of cooperation. *Science, 314*(5805), 1560–1563.

Nowak, M. A., & Sigmund, K. (1993). A strategy of win-stay, lose-shift that outperforms tit-for-tat in the Prisoner's dilemma game. *Nature, 364*(6432), 56–58.

Oestreich, A. M., & Tsiakas, I. (2015). Carbon emissions and stock returns: Evidence from the EU emissions trading scheme. *Journal of Banking & Finance, 58*, 294–308.

Orlitzky, M. (2011). Institutional logics in the study of organizations: The social construction of the relationship between corporate social and financial performance. *Business Ethics Quarterly, 21*(3), 409–444.

Orlitzky, M., Schmidt, F. L., & Rynes, S. L. (2003). Corporate social and financial performance: A meta-analysis. *Organization Studies, 24*(3), 403–441.

Peloza, J. (2009). The challenge of measuring financial impacts from investments in corporate social performance. *Journal of Management, 35*(6), 1518–1541.

Peloza, J., & Shang, J. (2010). How can corporate social responsibility activities create value for stakeholders? A systematic review. *Journal of the Academy of Marketing Science, 39*(1), 117–135.

Pierce, J. R., & Aguinis, H. (2013). The too-much-of-a-good-thing effect in management. *Journal of Management, 39*(2), 313–338.

Polbennikov, S., Desclée, A., Dynkin, L., & Maitra, A. (2016). ESG ratings and performance of corporate bonds. *The Journal of Fixed Income, 26*, 21–41.

Porter, M. E. (1979). How competitive forces shape strategy. *Harvard Business Review, 57*(2), 137–145.

Porter, M. E., & Kramer, M. R. (2006). The link between competitive advantage and corporate social responsibility. *Harvard Business Review, 84*(12), 78–92.

Preclaw, R., & Bakshi, A. (2015). *The cost of being green*. Barclays Research.

Ramiah, V., Martin, B., & Moosa, I. (2013). How does the stock market react to the announcement of green policies? *Journal of Banking & Finance, 37*, 1747–1758.

Revelli, C., & Viviani, J.-L. (2015). Financial performance of socially responsible investing (SRI): What have we learned? A meta-analysis. *Business Ethics: A European Review, 24*(2), 158–185.

Ridley, M., Chan, W.-C., & Edwards, C. (2016). *Global green bonds: The big long*. HSBC Global Research.

Rowley, T., & Berman, S. (2000). A brand new brand of corporate social performance. *Business & Society, 39*(4), 397–418.

Rubera, G., & Kirca, A. H. (2012). Firm innovativeness and its performance outcomes: A meta-analytic review and theoretical integration. *Journal of Marketing, 76*(3), 130–147.

Sandhu, S., Orlitzky, M., & Louche, C. (forthcoming). How nation-level background governance conditions shape the economic payoffs of corporate environmental performance. *Management Decision*. https://doi.org/10.1108/MD-11-2017-1191.

Sartzetakis, E. S. (2019). *Green bonds as an instrument to finance low carbon transition*. Working Paper, Bank of Greece.

Schmidt, F. L., & Oh, I.-S. (2013). Methods for second order meta-analysis and illustrative applications. *Organizational Behavior and Human Decision Processes, 121*(2), 204–218.

Schneeweiß, A. (2019). *Great expectations – Credibility and additionality of green bonds*. Bonn: Südwind e.V. – Institut für Ökonomie und Ökumene.

Servaes, H., & Tamayo, A. (2013). The impact of corporate social responsibility on firm value: The role of customer awareness. *Management Science, 59*(5), 1045–1061.

Sundaram, A. K., & Inkpen, A. C. (2004). The corporate objective revisited. *Organization Science, 15*(3), 350–363.

Tolliver, C., Keeley, A. R., & Managi, S. (2019). Green bonds for the Paris agreement and sustainable development goals. *Environmental Research Letters*, 14 (6).

Trikalinos, T. A., Churchill, R., Ferri, M., Leucht, S., Tuunainen, A., Wahlbeck, K., & Ioannidis, J. P. (2004). Effect sizes in cumulative meta-analyses of mental health randomized trials evolved over time. *Journal of Clinical Epidemiology, 57*(11), 1124–1130.

Trumpp, C., & Guenther, T. (2017). Too little or too much? Exploring U-shaped relationships between corporate environmental performance and corporate financial performance. *Business Strategy and the Environment, 26*(1), 49–68.

Waddock, S. A., & Graves, S. B. (1997). The corporate social performance-financial performance link. *Strategic Management Journal, 18*(4), 303–319.

Weber, J., & Gladstone, J. (2014). Rethinking the corporate financial-social performance relationship: Examining the complex, multistakeholder notion of corporate social performance. *Business and Society Review, 119*(3), 297–336.

Wernerfelt, B. (1984). A resource-based view of the firm. *Strategic Management Journal, 5*(2), 171–180.

Wulandari, F., Schäfer, D., Stephan, A., & Sun, C. (2018). *Liquidity risk and yield spreads of green bonds.* Discussion Paper, DIW Berlin.

Zerbib, O. D. (2019). The effect of pro-environmental preferences on bond prices: Evidence from green bonds. *Journal of Banking & Finance, 98*, 39–60.

CHAPTER 6

Institutional Initiatives to Foster Green Finance at EU Level

Vladimiro Marini

6.1 Introduction[1]

The European Union (EU) has a long-term commitment towards sustainability, innovation and a greener economy. Considering only the last 20 years, the *Lisbon strategy for growth and jobs*[2] in 2000 already mentioned the challenges linked to sustainable growth and the preservation of the environment. It was followed in 2007 by the European Commission's Communication on *An energy policy for Europe*,[3] which focused on energy efficiency and renewable energies (as well as internal market for gas and electricity, security of supply, international energy policy and energy technologies). Hence,

[1] The information contained in this chapter is updated as of June 2019. The reader can verify the presence of updates on the ongoing European Commission's initiatives for the development of sustainable finance here: https://ec.europa.eu/info/business-economy-euro/banking-and-finance/sustainable-finance_en#overview

[2] Presidency Conclusions (PC) of March 23rd–24th, 2000.

[3] European Commission (EC) (2007).

V. Marini (✉)
University of Rome Tor Vergata, Rome, Italy
e-mail: vladimiro.marini@uniroma2.it

M. Migliorelli, P. Dessertine (eds.), *The Rise of Green Finance in Europe*, Palgrave Studies in Impact Finance,
https://doi.org/10.1007/978-3-030-22510-0_6

following the *Paris Agreement* (UNFCCC 2015) and the United Nations' 2030 Agenda (UN 2015), both issued in 2015, in 2018 the European Commission further reinforced its pledge towards the environment with its *Long-term vision for a prosperous, modern, competitive and climate-neutral economy by 2050 – A Clean Planet for All*.[4] In this respect, it stated that "the purpose [...] is to create a vision and sense of direction, plan for it, and inspire as well as enable stakeholders, researchers, entrepreneurs and citizens alike to develop new and innovative industries, businesses and associated jobs".

The ultimate goal of the EU action in the last decades as concerns environmental issues can be hence summarised in an effort to integrate environmental and social transformations with other EU objectives such as competitiveness, high-quality jobs and economic growth. To this extent, the EU approach has constantly been to quantify environmental targets transparently to ease the process of tracking results and eventually allocate additional energies (Table 6.1 summarises some selected EU targets for

Table 6.1 Selected EU targets for the environment

Europe 2020[a]	*Europe 2030*[b]	*Europe 2050*[c]	*EU long-term strategy*[d]
20% reduction (at least) in greenhouse emissions compared to 1990 levels	40% reduction (at least) in greenhouse emissions compared to 1990 levels	80–95% reduction in greenhouse emissions compared to 1990 levels	Identified seven areas that require action: Energy efficiency Deployment of renewables Clean, safe and connected mobility Competitive industry and circular economy Infrastructure and interconnections Bio-economy and natural carbon sinks Carbon capture and storage to address remaining emissions
20% increase in the share of EU energy consumption coming from renewable sources	Minimum 27% share of renewable energy of 27%		
Improve energy efficiency to reduce the amount of primary energy used by 20% compared with projected levels	Minimum 27% improvement in energy efficiency		

Source: Author's elaboration

[a]European Commission (EC) (2010)
[b]European Commission (EC) (2014)
[c]European Commission (EC) (2011)
[d]European Commission (EC) (2018a)

[4]European Commission (EC) (2018a).

the environment set through the years). In such a framework, sustainable and green finance have been playing an increasing and distinctive role. As a matter of fact, financing the environmental transition requires a considerable amount of dedicated investments and resources, which need to be directed to targeted sectors and activities.

This chapter has as objective to describe how the different EU institutions have been supporting (and will support) the financing of the environmental transition by encouraging the development of sustainable and green finance. To do that, the rest of the chapter is organised as follows. Section 6.2 describes the pioneering activity of the European Investment Bank (EIB) in the issuance of green and sustainable financial securities. Sect. 6.3 considers the antecedents of the European Commission's *Action Plan for Financing Sustainable Growth*[5] (the Action Plan) issued in 2018, while Sect. 6.4 details this plan. Section 6.5 surveys the key (mainly ongoing) initiatives led by the European Commission as following the Action Plan, while Sect. 6.6 concludes the chapter by considering the main EU policies and instruments whose scope goes beyond Europe.

6.2 The Pioneering Role of the European Investment Bank in Developing Green and Sustainable Finance

The EIB was founded in 1958 as the bank of the EU and is among the largest and trusted financial institutions globally. Today, the EIB implements the EU policy with three macro-activities. First, lending, which is its traditional and indeed principal activity. Second, blending, which consists in the proposition of a variety of financial solutions to improve risk allocation and unlock additional capacity. Third, advising, namely providing financial and technical support to actual and potential borrowers. Given its reputation and risk-sharing instruments, the EIB's involvement often attracts new investors ("crowding-in"). The EIB's investment strategy focuses on innovation and skills, small and medium-sized enterprises (SMEs), infrastructure, climate and environment. In this respect, one single operation can serve multiple investment areas (e.g. an innovative environmental project of a European SME) and may require a combination of lending, blending and advising.

[5] European Commission (EC) (2018b).

In July 2007, the EIB conducted the first ever institutional green bond issuance (the same year as the EU Berlin Declaration on the occasion of the 50th anniversary of the signature of the Treaties of Rome[6]) in the framework of the EU energy policy (European Commission (EC) 2007). The first green bond was an AAA-rated *Climate Awareness Bond* (CAB) and was listed at the Luxembourg Stock Exchange. The proceeds supported projects on renewable energy and energy efficiency. More specifically, "ring-fencing of proceeds" for risk hedging and disclosure, "reporting on use of proceeds" for transparency, "disbursements to new energy efficiency and renewable energy loans only" for credibility were the key conditions attached to the issuance. In technical terms, the value components offered to investors were the EIB's credit quality (exposure to EIB lending, not to projects), capital protection (100% of the nominal investment) and a guaranteed minimum return (5% of the nominal and 80% participation in the potential positive performance of a European equity index over a period of five years[7]). Since its first issuance, the EIB became the main institutional issuer of green bonds, including sustainable bonds, named *Sustainability Awareness Bonds* (SAB): as of February 2019, almost EUR 24 billion were issued across 11 currencies,[8] and related proceeds have been supporting 160 renewable energy, energy efficiency and other sustainable projects globally (Table 6.2 presents the EIB historical issuance of CAB and SAB). Regarding in particular SAB, they have contributed to leverage the experience gained with CAB, and they extend the commitment of the EIB beyond climate action towards environmental and social matters, thus implementing EU sustainability strategy more extensively. In this respect, proceeds are to be linked to objectives rather than activities, and those objectives are open-ended. SAB proceeds typically fund EIB's lending to projects on the sustainable use and protection of water and marine resources, transition to a circular economy, waste prevention and recycling, pollution prevention and control, and protection of healthy ecosystems.

[6] European Union (EU) (2007).

[7] This index was the FTSE4Good European Environmental Leaders40 Index. Forty European companies with the highest environmental rating, adjusted for market capitalisation were included in the index, recalculated twice a year. Non-green industries were also included to incentivise them towards sustainability and to compensate the volatility of the green component.

[8] EUR, USD, GBP, SEK, CAD, ZAR, CHF, AUD, JPY, TRY, and BRL.

Table 6.2 EIB issuances of CAB and SAB

Type	*Launch date*	*Maturity*	*Coupon*	*Bond currency*	*Amount issued (m)*	*Amount issued in EUR (m) equivalent*
CAB	02/07/2007	28/06/2012	Not applicable	EUR	600	600
CAB	03/11/2009	17/02/2015	2.95%	SEK	2400	241
CAB	03/11/2009	17/02/2015	FRN*	SEK	550	52
CAB	28/01/2010	16/03/2016	0.5%	BRL	60	15
CAB	23/02/2010	16/03/2015	8%	BRL	303	123
CAB	24/02/2010	17/03/2014	7.43%	ZAR	86	8
CAB	24/02/2010	17/03/2014	4.83%	AUD	18	12
CAB	28/04/2010	29/05/2013	6.68%	ZAR	1375	139
CAB	28/04/2010	24/05/2012	4.27%	AUD	231	161
CAB	25/10/2010	21/11/2013	6.62%	TRY	170	86
CAB	10/04/2012	23/04/2019	3%	SEK	3750	428
CAB	31/10/2012	13/11/2023	2.75%	SEK	2175	237
CAB	11/07/2013	15/11/2019	1.375%	EUR	3000	3000
CAB	17/07/2013	24/07/2020	FRN*	SEK	1800	203
CAB	07/11/2013	15/09/2017	6.75%	ZAR	2300	164
CAB	08/01/2014	04/02/2025	1.625%	CHF	350	283
CAB	26/02/2014	12/03/2018	7.75%	ZAR	500	35
CAB	03/03/2014	25/03/2039	PRDC**	JPY	5000	36
CAB	26/03/2014	07/03/2020	2.25%	GBP	1800	2375
CAB	03/09/2014	13/11/2026	1.25%	EUR	1800	1800
CAB	08/10/2014	15/10/2024	2.5%	USD	1000	794
CAB	26/02/2015	27/03/2019	8.50%	TRY	275	84
CAB	20/08/2015	15/11/2023	0.5%	EUR	1900	1900
CAB	27/10/2015	05/11/2020	9.25%	CAD	500	342
CAB	13/11/2015	18/05/2029	1.75%	EUR	500	500
CAB	11/01/2016	20/01/2021	0.625%	SEK	1000	108
CAB	06/04/2016	13/04/2026	2.125%	USD	1500	1319
CAB	09/09/2016	16/09/2021	1.125%	CAD	500	343
CAB	28/09/2016	13/11/2037	0.5%	EUR	1250	1250
CAB	11/01/2017	19/07/2022	0.5%	SEK	3000	314
CAB	21/02/2017	02/03/2027	1.50%	SEK	2500	260
CAB	17/05/2017	24/05/2027	2.375%	USD	1500	1346
CAB	28/06/2017	15/11/2047	1.5%	EUR	1250	1250
CAB	25/07/2017	03/02/2028	3.3%	AUD	1250	814
CAB	03/01/2018	12/01/2023	2.7%	AUD	750	488
CAB	10/01/2018	18/01/2023	2.375%	CAD	700	468
CAB	17/01/2018	30/01/2025	0.875%	SEK	1800	183
CAB	18/04/2018	13/06/2025	2.875%	USD	1500	1211
CAB	22/05/2018	15/11/2032	1.125%	EUR	500	500
SAB	06/09/2018	15/05/2026	0.375%	EUR	500	500
Total issued						23,972

Source: Author's elaboration on data EIB (EIB's first Climate and Sustainable Awareness Bonds Newsletter 2019, page 4). Data as of February 2019. Launch date refers to the date the line was originally launched. Amount issued includes potential taps. *FRN stays for floating-rate note. **PRDC stands for power reverse dual-currency note

Besides green bonds and sustainable bonds, the EIB has been increasingly operating on green lending, blending and advising within its operational plans and its broader mandate to implement the EU policy. Moreover, it has been playing a leading role in the international debate (especially with Chinese institutions[9]) aimed at formulating globally shared principles, on which national and international green programmes may be grounded.

6.3 Towards an EU Action Plan for Sustainable Finance: The High-Level Expert Group on Sustainable Finance

One of the cornerstone initiatives of the EU to foster sustainable finance was established in December 2016 by the European Commission, namely the High-Level Expert Group (HLEG) on sustainable finance.[10] The HLEG had the role to define the basis for a comprehensive European strategy on sustainable finance as part of the EU commitment to a more sustainable economy. Accordingly, the HLEG was mandated to provide a roadmap containing practical policy recommendations on how to orient private capital towards sustainable investing and growth, assess and manage environmental risks and supervise risk takers, and integrate sustainability concerns into the EU policy framework. The HLEG was composed of senior experts from civil society, financial services industry, academia and European institutions with strong expertise on funding, project evaluation and lending (the EIB has been required by the European Commission to assist the HLEG as observer and adviser). As a result of this initiative, the HLEG published an interim report in July 2017,[11] launched a public consultation to gather feedback and then published its final report in January 2018.[12] In the latter it was clarified that "the ultimate test of the HLEG will not just be the degree to which its specific recommendations are adopted, but the extent to which sustainable finance becomes a permanent feature of European markets and policy-making" (HLEG 2018).

In its final recommendations, the HLEG first emphasised the priority of establishing an EU taxonomy of sustainable activities. In this respect, the

[9] See also Sect. 6.6.1.

[10] With Communication "*Capital Markets Union – Accelerating reform*", European Commission (EC) (2016a).

[11] High-Level Expert Group on Sustainable Finance (HLEG) (2017).

[12] High-Level Expert Group on Sustainable Finance (HLEG) (2018).

Table 6.3 Possible uses of an EU taxonomy

#	Possible uses of the EU sustainability Taxonomy
1	Measuring financial flows towards sustainable development.
2	Identifying assets that qualify for financing under European sustainable funding mechanisms.
3	Providing a starting point for standard-setters and product developers, for example, green bonds or research/index providers.
4	Allowing investors to understand the green/sustainable exposure of their portfolios.
5	Providing breakdowns of companies according to various sustainability categories.
6	Supporting investor engagement with companies around their business models and transition plans.
7	Promoting disclosure in line with the Financial Stability Board's Task Force on Climate-related Financial Disclosures (Financial Stability Board (FSB) 2017) and informing developments of the Non-Financial Reporting Directive.

Source: Author's elaboration on High-Level Expert Group on Sustainable Finance (HLEG) (2018)

lack of clarity about definitions attached to sustainable finance and the perceived risk of greenwashing were mentioned as potentially main reasons for the sustainability investment gap and the potential greater volatility of green investments.[13] Table 6.3 summarises the main instrumental use on an EU taxonomy of sustainable activities.

In addition, the HLEG concluded that private capital could be oriented towards green investing by designing new financial solutions based on environmental impact/contribution assessment, and including a strong technical advisory component in supporting green projects. This would facilitate the integration of green finance in financial systems, whose stability indeed remains a priority (HLEG 2018). Beyond that, the HLEG covered several other themes, including recommendations on the need for an improved contribution of the financial system to sustainable and inclusive growth: this would be reached by a greater long-term orientation and the strengthening of financial stability by incorporating environmental, social and governance (ESG) factors into decision-making. A summary of the HLEG final report is given in Table 6.4.

[13] The HLEG did not furnish the taxonomy but suggested the framework on which it should be grounded.

Table 6.4 Summary of the HLEG final report

Priority actions. ***The HLEG recommends:***
Establishing an EU sustainability Taxonomy, starting with climate mitigation, to define areas where investments are needed most
Clarifying investor duties to extend the time horizons of investment and bring greater focus on environmental, social and governance (ESG) factors into investment decisions
Upgrading disclosures to make sustainability opportunities and risks transparent
Enabling retail investors to invest in sustainable finance opportunities
Developing official European sustainability standards for some financial assets, starting with green bonds
Establishing "Sustainable Infrastructure Europe" to deploy development capacity in EU member states for infrastructure necessary for a more sustainable economy
Integrating sustainability firmly in the governance of financial institutions as well as in financial supervision
Cross-cutting recommendations. ***The HLEG advises the EU:***
To confront short-termism in financial markets so as to reduce its negative impact on long-term corporate investment and development
To consider ways to empower citizens to engage with sustainable finance
To monitor investment plans and delivery through a dedicated EU observatory on sustainable finance
To improve financial market benchmark transparency and guidance
To ensure that EU accounting rules do not unduly discourage long-term investment
To establish a "Think Sustainability First" principle at the heart of EU policy-making and to drive sustainable finance at the global level
Recommendations for specific sectors of the financial system. ***Their purpose is:***
To promote real economy and sustainability lending in the banking sector
To enable insurance companies to have a stronger role in equity, long-term and infrastructure investments
To ensure that asset managers, pension funds and investment consultants grasp the sustainability preferences of their clients
To ensure that credit rating agencies lengthen the time horizon of risk analysis and disclose how they consider ESG factors
To have listing authorities promote disclosure of ESG information
To obtain better long-term research by investment banks
Other social and environmental challenges
Supporting the growth of social enterprises and the financing of social-related projects
Revaluing natural and environmental capital in economic and financial decisions
Reorienting agriculture to a way that is more sustainable for the economy, the environment and public health

Source: Author's elaboration on High-Level Expert Group on Sustainable Finance (HLEG) (2018)

6.4 The European Commission's Action Plan for a Greener and Cleaner Economy

Building on HLEG's work, the European Commission released in March 2018 its *Action Plan for Financing Sustainable Growth*.[14] The Action Plan had the merit to stress the crucial role of finance in reaching a sustainable economy and to identify concrete measures to foster sustainable finance. In this respect, the Action Plan specified the EU objectives to:

- reorient capital flows towards sustainable investment in order to achieve sustainable and inclusive growth;
- manage financial risks stemming from climate change, resource depletion, environmental degradation and social issues;
- foster transparency and long-termism in financial and economic activity.

Its motivation was to identify concrete actions able to systematically integrate sustainability into the EU policy, without threatening financial stability. In this respect, the Action Plan also aimed at "*not increasing the overall regulatory burden and complexity, given that the ultimate purpose is to facilitate more investment*" (EC 2018b). To this extent, ten different actions were identified. At the time of the redaction of this contribution,[15] the implementation of the Action Plan is still ongoing (a detailed overview on the specific deadlines for the first initiatives linked to each action is given in Annex 6.1), and little concrete results can be hence already reported and discussed. In this background, this section provides a summary on the ten actions shaping the Action Plan (they are reported hereafter with respect to the objective they intend to serve) and explains their rationale. The next section gives more information on some of the key ongoing initiatives following the plan.[16]

[14] European Commission (EC) (2018b).

[15] May 2019.

[16] To support the Action Plan, in May 2018 the Commission also adopted a package of measures (European Commission (EC) 2018c, d, e) implementing key actions announced. The package included:

- a proposal for a regulation on the establishment of a framework to facilitate sustainable investment, setting up the conditions and the framework to gradually create a unified classification system (taxonomy) on what can be considered an environmentally sustainable economic activity (see Sects. 6.4.1 "Action 1", and 6.5.1 on this topic);

6.4.1 Reorienting Capital Flows Towards Sustainable Investment in Order to Achieve Sustainable and Inclusive Growth

As regards the first objective of the Action Plan, reorienting capital flows towards sustainable investment in order to achieve sustainable and inclusive growth, the Commission estimated on the 2030 horizon an investment gap per year of EUR 180 billion in sustainable investing,[17] and of EUR 270 billion in transport, resources management and infrastructures.[18] In an attempt to bridge this gap and trigger the necessary investments, five actions have been identified.

Action 1: Establishing an EU Classification System for Sustainable Activities (Taxonomy)

Defining what is sustainable and what is not sustainable is indeed recognised as a fundamental basis for the development of sustainable finance. The lack of shared definitions (based on screening criteria, thresholds and metrics of sustainability-related activities) should indeed be considered one of the main issue to overcome in order to provide all stakeholders with the necessary enabling investment framework for sustainable finance. For this reason, the Action Plan sets the definition of a taxonomy of sustainable activities as one of the priorities. Nevertheless, the definition of a taxonomy is not a straightforward exercise as it implies a thorough analysis of the impact of each economic activity on the different dimensions of the sustainability landscape, including but not limited to environmental, social and governance issues.[19]

- a proposal for a regulation on disclosures relating to sustainable investments and sustainability risks and amending Directive (EU) 2016/2341 to introduce disclosure obligations on how institutional investors and asset managers integrate environmental, social, and governance (ESG) factors in their risk processes (see Sects. 6.4.2 "Action 7", 6.4.3 "Action 9", and 6.5.3 on this topic);
- a proposal for a regulation amending the benchmark regulation to create a new category of benchmarks comprising low-carbon and positive carbon impact benchmarks, which will provide investors with better information on the carbon footprint of their investments (see sect. 6.4.1 "Action 5" on this topic).

[17] Based on the Commission's Impact Assessment for the amendment of the Energy Efficiency Directive (EC 2016b).

[18] See also the EIB Operational Plan 2017–2019 of December 2016 (EIB 2016).

[19] See also Sect. 6.5.1, and for a more detailed analysis of the taxonomies developed in the field of green finance, see Chap. 2.

Action 2: Creating Standards and Labels for Green Financial Products

Building on the taxonomy, an identification system to allow investors to consistently identify green financial products is also foreseen to strengthen investors' confidence in the market. In this respect, the European Commission will specify, *inter alia*, the specific contents of the prospectus based on the existing applicable regulation.[20] This framework is expected to foster in particular green bonds, whose volume is today only about 1% of the bond market[21] but expected to grow further.

Action 3: Fostering Investment in Sustainable Projects

The Action Plan foresees the deployment of specific financing instruments at the EU level. This action may result in the further development of the European Fund for Sustainable Development (EFSD) and the European Fund for Strategic Investments (EFSI). In addition, following the Action Plan, the Commission may establish a single EU fund for sustainable investment to enhance the interaction between investors, beneficiaries, the European Commission and implementing partners (EC 2018b). In this respect, despite large-scale projects catalyse more capital and have positive externalities, supporting smaller projects also has been considered an EU priority that can contribute to speed-up the transition towards a sustainable economy.

Action 4: Incorporating Sustainability When Providing Financial Advice

The second Markets in Financial Instruments Directive (MiFID II) and the Insurance Distribution Directive (IDD) do not include sustainability preferences among the requisites to identify a "suitable" product to meet clients, (financial) needs. This may indeed hamper the effective demand of sustainable financial products from retail investors to be conveyed in the financial markets. The European Commission hence invited ESMA[22] and EIOPA[23] to provide solutions to bridge this gap, and will hence gather feedback about potential amendments to MiFID II and IDD.

[20] Regulation (EU) 2017/1129 on the prospectus to be published when securities are offered to the public or admitted to trading on a regulated market (and repealing Directive 2003/71/EC) (hereinafter EU 2017).

[21] Fitch Ratings (2017), "Green Bonds – Fitch Ratings and Market Overview".

[22] European Securities Markets Authority.

[23] European Insurance and Occupational Pensions Authority.

Action 5: Developing Sustainability Benchmarks

Sustainability benchmarks would ease a responsible asset allocation as they have the potential to drive market demand. Nevertheless, the opacity of current ESG benchmarks in the market hampers their credibility and diffusion. In this respect, the solution proposed in the Action Plan consists in aligning benchmarks to qualified policy sustainability targets. Following the Action Plan, the European Commission has already tabled a proposal for amending the benchmark regulation aiming at "*increasing their transparency and effectiveness so that the resulting benchmark portfolio has less emissions compared to [...] a standard benchmark*".

6.4.2 Managing Financial Risks Stemming from Climate Change, Resource Depletion, Environmental Degradation and Social Issues

With respect to the second objective of the Action Plan, that is managing financial risks stemming from climate change, resource depletion, environmental degradation and social issues, three actions have been identified. These actions are expected to contribute to face the financial consequences of the observed weather-related losses rise (by +86% in 2007–2016 and amounting to EUR 117 billion in 2016[24]) and of the increasing exposure of financial institutions to climate-related risks (as 50% of European banks can be considered to be effectively exposed to this type of risks[25]).

Action 6: Better Integrating Sustainability in Ratings and Market Research

Market research providers and rating agencies are already considering environmental, social and governance (ESG) factors in their assessment activities. However, as for sustainability benchmarks, opacity (as concerns quality of information and methodologies) is still a main issue to overcome. In this respect, following the Action Plan, ESMA will promote solutions in coordination with the European Commission, which may amend the Credit Rating Agency Regulation. Better integrating sustainability considerations in existing rating system can indeed represent an effective tool to manage the financial risks stemming from climate change. In this respect, credit rating and market research organisations should in

[24] Source: Lancet (2017).

[25] Battiston et al. (2017).

particular correct their typical bias towards larger players, whose disclosures are of higher quality: this could further help to clarify the role of smaller players in sustainability issues.

Action 7: Clarifying Institutional Investors' and Asset Managers' Duties
Sustainability cannot be fully addressed within the traditional interpretation of the concept of *best interest*. The Action Plan aims at explicitly requiring institutional investors and asset managers to integrate sustainability considerations in their investment decision-making process and increase transparency towards end investors on how to integrate sustainability factors in their investment decisions. In this respect, the European Commission's *proposal for a regulation on disclosures relating to sustainable investments and sustainability risks* of May 2018 (European Commission (EC) 2018d), following the Action Plan, already indicates, *inter alia*, that financial market participants and insurance intermediaries should adopt ESG factors and provide sustainability reports in written form.

Action 8: Incorporating Sustainability in Prudential Requirements
The last amendment of Basel III of December 2017[26] did not take into account sustainability-related matters systematically. Once the taxonomy will be published (as a basis for defining which are the sustainable activities), the European Commission will explore the feasibility of the inclusion of risks associated with climate and other environmental factors in financial institutions' risk management policies and the potential calibration of capital requirements. In this respect, relevant EU laws may eventually integrate ESG factors in the concept of financial stability.

6.4.3 Fostering Transparency and Long-Termism in Financial and Economic Activity

Finally, in order to reach the third objective of the Action Plan, namely fostering transparency and long-termism in financial and economic activity, two additional actions have been identified. In this respect, the final goal is to induce market investors to consider sustainable finance as a reference asset class in the long-term, in this way reducing short-term market pressure based exclusively on financial returns.

[26]The last amendment of Basel III is also known as Basel IV by analysts and practitioners.

Action 9: Strengthening Sustainability Disclosure and Accounting Rule-Making

Corporate sustainability reporting is addressed by the Directive on the Disclosure of Non-Financial Information (NFI Directive). In order to improve the quality and the quantity of climate-related information provided by businesses and following the Action Plan, in 2018 the European Commission has established a Corporate Reporting Lab[27] as part of the European Financial Reporting Advisory Group (EFRAG)[28] to promote innovation and the development of best practices in corporate reporting, such as environmental accounting. Relatedly, the European Parliament and the European Commission raised concerns about the application of the IFRS 9 as concerns sustainability matters and its fit with long-termism.

Action 10: Fostering Sustainable Corporate Governance and Attenuating Short-Termism in Capital Markets

Corporate governance can play a fundamental role to reduce short-term pressures in internal processes and strategies as to reduce exposure to environmental risks. However, current efforts towards sustainability appear to be insufficient as regards their scope, effectiveness and comparability. Accordingly, in February 2019, the European Commission has mandated EIOPA, EBA[29] and ESMA to collect evidence to tackle those issues (while avoiding trade-offs with other EU policy targets).

6.5 Following the Action Plan: Main Ongoing Policy Initiatives to Develop Sustainable Finance in Europe

6.5.1 *The Establishment of a Shared, Universal Taxonomy*

The importance of the development of an EU taxonomy for sustainable activities had been already highlighted by the HLEG and a specific action was included in the Action Plan. In this respect, the European Commission's *proposal for a regulation on the establishment of a framework to facilitate*

[27] "*The first project of the European Lab* [...] *will assess* [...] *current and potential use of climate-related information*". Source: Corporate Reporting Lab website.

[28] The EFRAG was established in 2001 and its Member Organisations have expertise and interest in developing IFRS.

[29] European Banking Authority.

sustainable investment of May 2018 (European Commission (EC) 2018c) complements the Action Plan by providing a framework for the development of the taxonomy. The proposal states that an economic activity is *sustainable* when belonging to (i) climate change mitigation; (ii) climate change adaptation; (iii) sustainable use and protection of water and marine resources; (iv) transition to a circular economy, waste prevention and recycling; (v) pollution prevention and control and (vi) protection of healthy ecosystems. In addition, the economic activity should *not significantly harm* other environmental objectives (EC 2018c). Besides its relevance, the highly technical nature of the taxonomy has induced the European Commission to prioritise the work on climate change mitigation and climate change adaptation. Only in a second phase a more exhaustive version of the taxonomy will be released. In addition, its integration in the EU law will be gradual.

Practically, the definition of the taxonomy requires the establishment of quantitative thresholds and metrics for identifying relevant assets and activities (e.g. the maximum level of greenhouse gas emissions to consider an activity as contributing to climate change mitigation). In this respect, in July 2018, the European Commission established a Technical Expert Group (TEG) on sustainable finance, set to operate (at least) until June 2019.[30] The TEG was mandated to improve the functioning of mainstream capital markets in favour of sustainability by also contributing to establish the EU taxonomy.[31] The TEG is composed of members from academia, business and financial sector, plus observers from international bodies. The TEG released its technical report in June 2019 covering climate change mitigation and climate change adaptation activities (TEG 2019b). The recommendations in the report provide the basis for the future EU taxonomy.[32,33]

[30] With a possible extension until year-end 2019.

[31] In more detail, the TEG objective is to assist notably in the development of a unified classification system for sustainable economic activities, an EU green bond standard, methodologies for low-carbon indices, and metrics for climate-related disclosure.

[32] The report issued in June 2019 has been set open for feedback until September 2019. Feedback received will be incorporated into a report submitted to the Commission in late 2019. The establishment of the EU taxonomy for climate change mitigation and climate adaptation activities is expected in 2020 (*this chapter has been finalised in July 2019*).

[33] For further details on the role of taxonomies in green finance, see Chap. 2.

6.5.2 *Definition of EU Sustainable Labels in the Financial Sector*

Current volumes of finance towards sustainable and green projects largely depend on institutional credibility (as in the case of the *Climate Awareness Bonds* issued by the EIB). In this respect, the EU taxonomy will likely increase the mid-term consistency and reliability of project identification and green instruments issuance also for non-institutional issuers (including project promoters). This would probably incentivise the growth of green financial centres and ultimately contribute to reorient private capital towards green investments (EC 2018b). However, in particular, in the case of non-institutional issuers, retail investors could still perceive insufficient credibility and opacity about the financial instruments associated with green projects. Accordingly, and building on the EU taxonomy, the European Commission is working on the definition of rules for voluntary standards and labels for green financial products. This would enable market credibly and help investors to select the financial instruments that orient funds to green projects, assets or business activities. In other terms, investors could easily express their preferences for sustainable investing within the broader spectrum of available investment opportunities: this would foster the choice of green finance in current investment practices (e.g. by integrating standards and labels in comparison tools or financial planning services[34]) rather than considering sustainable investing as a stand-alone asset class. In this regard, the Commission will also explore how the voluntary labelling scheme for financial products could be compatible with the EU Ecolabel Regulation.[35] Notably, the voluntary standards approach

[34] Under the European Commission's Consumer Financial Services Action Plan of 2017 (EC 2017a).

[35] The EU Ecolabel is a sector-based voluntary scheme according to which producers, importers, and retailers can credibly define their products and services as environmentally friendly. The Regulation on Ecolabel started in 1992: its objective is "*to promote products with a reduced environmental impact during their entire life cycle and to provide consumers with accurate, non-deceptive, science-based information on the environmental impacts of products*". EU Ecolabels are governed by:

- The European Union Ecolabelling Board (EUEB) responsible of the development, revision, and implementation in close collaboration with the Commission
- The Commission, ultimately responsible for the functioning of the Scheme (EU Ecolabel updates are named "Commission decisions")
- Competent Bodies (CBs), independent organisations designated by the EEA responsible for national transposition and implementation, including the verification of applicants' proposals and the update of User Manuals

adopted by the European Commission is aimed at rendering the new schemes more inclusive in its early phase. After green labels, the Commission will extend the classification system to social matters, building in particular on the experience of European Social Entrepreneurship Funds. Standard and labels for sustainable financial products would complement the other tools aimed at reducing opacity to retail investors, namely, sustainability rating, benchmarking and asset management (EC 2018b).

6.5.3 *Sustainable Reporting and Climate-Related Financial Disclosure*

The Action Plan also addresses the issue of transparency. According to the Directive on the Disclosure of Non-financial Information,[36] member states can already require third-party verifications, and companies are encouraged to share best practices on their disclosure standards. Also, public-interest entities[37] have to provide detailed report about policies, outcomes and risks related to at least environmental protection, social responsibility and treatment of employees, respect for human rights, anti-corruption and bribery, and diversity on company boards.[38] Companies have to disclose relevant, useful and comparable information according to existing guidelines.[39]

- Stakeholders, namely industry representatives and service providers, including SMEs, trade unions, retailers, importers, environmental protection groups and consumer organisations

The rationale is identifying the phases of a product/service life cycle in which its environmental impact is significant as to implement necessary technical remedies. The impact could be at consumption, thus requiring energy efficiency remedies, rather than process-related solutions. Technical discussions take place at Ad Hoc Working Group meetings and their conclusions are grounded on EUEB guidance. EU Ecolabels can be used within Public Procurement (15–20% of EU GDP), namely a voluntary "*process whereby public authorities seek to procure goods, services and works with a reduced environmental impact throughout their life cycle*". However, Public Procurers are not allowed to require products with EU Ecolabels, but at most that some Ecolabel criteria are met. Otherwise, the activity of Green Public Procurement would have been based on EU Ecolabel only, rather than a combination among existing law and the EU Ecolabel scheme.

[36] Directive 2014/95/EU.

[37] Listed companies, banks, and insurance companies with more than 500 employees in 2017 (i.e. 6000 out of the EU 40,000 large companies are captured by this criterion).

[38] "*Member States* [can] *exempt undertakings when a separate report corresponding to the same financial year and covering the same content is provided*". Source: 2014/95/EU.

[39] For example, OECD Guidelines for Multinational Enterprises, EMAS or ISO 26000. This variety could require further harmonisation.

European Commission's Guidelines on Non-financial Reporting

The European Commission released a first version of non-binding guidelines about non-financial reporting[40] in June 2017, drawing on existing frameworks, the HLEG's inputs and stakeholders' feedbacks. Another reference in this field is Financial Stability Board (FSB) (2017). The final version of the guidelines is expected by June 2019.[41] These guidelines are aimed at promoting best practices and they are not sector-specific. The principle of *materiality* is recalled and should lead information disclosure. Information disclosure complies with the *materiality* principle when "*its omission or misstatement could reasonably be expected to influence decisions that users make on the basis of the financial statements of the undertaking*". In addition, "*the materiality of individual items shall be assessed in the context of other similar items*" (Directive 2013/34/EU, Art. 2, 16). Materiality assessment should benefit from corporate governance quality, reliable evidence, stakeholder engagement, independent verification, adapted public policy and regulation. In case of commercially sensitive information, disclosure can be provided in broader terms (EC 2017b).

TEG Report on Climate-Related Disclosure

The TEG published a Report on Climate-related Disclosures[42] as part of its mandate and following feedbacks from stakeholders. In TEG's view, climate-related disclosure should be open-ended while enabling horizontal and vertical comparability, namely, among companies and of a company over time (TEG 2019a). In Table 6.5 the benefits of environmental reporting according to TEG are reported. The TEG classifies three types of possible disclosures (TEG 2019a): those that all companies should disclose (type 1), those that companies should consider disclosing (type 2) and those that companies may consider disclosing (type 3).[43] In this respect, key performance indicators (KPIs) should be tailored to official goals to track the contribution to sustainability goals. Nevertheless, con-

[40] European Commission (EC) (2017b).

[41] "*TCFD is the first framework with the potential to become a 'new normal' of climate disclosure*" [...] and "*will lead more efficient allocation of capital, and help smooth the transition to a more sustainable economy*". Sources: High-Level Expert Group on Sustainable Finance (HLEG) (2018) and Financial Stability Board (FSB) (2017).

[42] Technical Expert Group on Sustainable Finance (TEG) (2019a).

[43] In particular, in terms of additional or innovative disclosures that enhance transparency.

Table 6.5 Main benefits of environmental reporting

Listed companies
Better understanding of the company's exposure to climate risks
Improved credit standing
Banks
Better understanding of exposures to climate-related risks
Evidence of their skills/expertise in climate-related transactions
Insurance companies
Evidence of risk control for financial regulators (stress testing)
Better understanding/management of climate-related risks and opportunities

Source: Author's elaboration on Technical Expert Group on Sustainable Finance (TEG) (2019a, b)

cerns can arise for some widely used KPIs, such as the Green Bond Ratio[44] and the Green Debt Ratio,[45] because of their simplicity and lack of depth. Complementing information based on best practices and overall frameworks is hence encouraged. The TEG also published a summary of comments received from qualified respondents in February 2019, mostly organisations and companies (80.6%). Respondents called for lower levels of granularity in the early phase, clearer links among targets and KPIs, more sector-specific guidance and more clarity about materiality.

6.5.4 *The Launch of Financial Support Initiatives*

The European Fund for Strategic Investments (EFSI)

As a response to declining investment levels in the EU, the EIB and the European Commission have launched the European Fund for Strategic Investments (EFSI) in November 2014, with the aim to generate EUR 315 billion of new investments by mid-2018. Given its relative success, in September 2016, the EFSI was extended to generate EUR 500 billion by 2020.[46] Still, the investment pattern was rather positive, especially since 2018.[47] In terms of investment targets, the EFSI operates as a trust-fund with two main goals: fostering growth, especially through the Infrastructure

[44] Green bonds outstanding/all bonds outstanding.

[45] Green bonds or equivalents outstanding/debt outstanding.

[46] The EFSI *"is a EUR 26 billion guarantee from the EU budget, complemented by a EUR 7.5 billion allocation of the EIB"*. Source: EIB and EFSI official websites.

[47] European Commission (EC) (2018f).

Table 6.6 Potential beneficiaries and intermediaries of EFSI funds

Private-sector entities	Large businesses, special purpose vehicles and medium-sized companies with up to 3000 employees (also called mid-caps) can benefit from project loans[a] or loans to finance research and innovation.[b] Mid-caps and small companies of less than 250 staff can also apply for growth finance or intermediated lending provided by financial partners and may benefit from EIF's intermediated equity or guarantee product.
Public-sector entities	Local authorities, public-sector companies or other government-related entities may benefit from project loans or loans to finance research and innovation. Smaller projects may also be financed through EIB's intermediated lending provided by partner institutions.
Banks, national promotional banks or other financial institutions	The EIB provides loans to institutions acting as financial intermediaries by financing small investments by SMEs (less than 250 employees), mid-caps (250 to 3000 employees) or local authorities. Financial intermediaries may also benefit from EIB portfolio guarantee instruments[c] as well as EIF products.
Funds and any other form of investment vehicles	EFSI can support investments in equity, hybrid or debt funds with a focused investment strategy addressing EU priorities and through the EIF funds focusing on SMEs.
Investment platforms	The EIB may co-invest or provide co-financing to eligible projects alongside third parties, such as investment platforms, either under fully delegated structures or with active EIB involvement into the due diligence and structuring process.

Source: Author's elaboration on information retrieved from the EIB website (https://www.eib.org/en/efsi/how-does-a-project-get-efsi-financing/index.htm)

[a]Minimum EUR 25 million, fixed, floating, revisable or convertible interest rate, loan repayment on a semi-annual or annual basis

[b]Up to 10 years, minimum EUR 25 million, senior/subordinated, secured/unsecured. Pricing reflects credit risk and EIB funding

[c]Senior or mezzanine. The pricing reflects the credit risk profile and EIB funding availability

and Innovation Window, and supporting SMEs, in particular through the SME Window.[48] By investing in risky projects (or tranches), the EFSI can also complements other financial schemes and may attract new investors in this way.[49] Notably, the EFSI credit standing is AAA. Table 6.6 summarises the potential borrowers and related financial solutions within the EFSI.

[48] EUR 240 and 75 billion mobilised, respectively.

[49] For an example as funding as EFSI could complement other financial schemes, see Migliorelli and Dessertine (2018).

The EFSI has a specific governance structure. A Steering Board, in charge of strategic orientation, risk profile management, operating policies and procedures. An Investment Committee, monitoring of EFSI's compliance to relevant regulation (members of the Investment Committee are expert in project structuring and financing, economic analysis in research and innovation, transport, renewable energy, education and health). A Managing Director, in charge of ordinary management, including chairing executive meetings and providing quarterly reports to the Steering Board. In addition, a European Investment Advisory Hub (EIAH) operates in close collaboration with the EFSI. It represents a "one-stop-shop" for technical assistance established in September 2015,[50] and it serves three types of audience: project promoters, investors and public managing authorities. It enhances the capacity to identify, prepare, design, structure, implement and monitor projects, as well as to exploit EU programmes.

In terms of its policy, the EFSI operates in the following sectors: strategic infrastructure including digital, transport and energy; education, research, development and innovation; renewable energy and resource efficiency; support for SMEs. Even though renewable energy and energy efficiency are not exhaustive with respect to sustainability, their inclusion in a broader strategy improves the integration of green investing in the whole EU policy. Moreover, EFSI Regulation[51] includes a flexible list of eligible sectors, and projects are selected case by case (also based on the EIB's due diligence and loan application procedures).[52] Projects have to provide additionality, that is, addressing projects that otherwise would not have been carried out. Following the Action Plan, the EFSI will further reinforce its focus on sustainable projects with at least 40% of its resources financing infrastructures and innovation to support climate change.

The European Investment Bank Implementing Tools

As previously mentioned, the EIB implements the EU policy by three macro-activities: lending, blending and advising. As concerns lending, since 2010 the EIB has a formal target to devote at least 20% of its annual

[50] In its first year of operation, the EIAH had received 239 requests (two-thirds from the private sector).

[51] European Union (EU) (2015).

[52] European Investment Bank (EIB) (2017).

lending to green projects (25% from 2011[53] and 35% by 2020[54] including developing countries). In this respect, the EIB provides financial and technical support to projects belonging to the environmental objectives and focuses on increasing the technical component of green lending rather than the mere volume. The EIB's green lending activity received a renewed impulse since 2015 following the key international developments (in particular, the *Paris Agreement*[55]). The activity of blending (consisting of combining tailored, complex and innovative financial solutions[56] with other sources of finance) includes the management of some financing schemes and platforms having a general scope (e.g. European Fund for Sustainable Development), while others are more specific (e.g. Private Finance for Energy Efficiency). Environmental considerations are increasingly included in the selection of the project to be financed. Finally, as concerns the advising activity, it is directed to help project developers that do not have the needed in-house experience, expertise and knowledge to develop a sustainable project. In particular, adaptation to climate change is often a preventive action that could not produce any immediate revenue, implying the need of specialised competences to set well-functioning and tailored financing instruments. To overcome these barriers, the EIB has launched stand-alone advisory services,[57] which can include market studies, programme structuring, energy audits, project preparation, implementation support, planning decisions, systems resilience, technology selection, project design and capacity building. Eventually, raising awareness about risks and knowledge gaps would inform potential project developers about how the project is vulnerable to environmental risks, and identify solutions to increase resilience. The environmental risk assessment is integrated in the process, rather than consisting of a separated phase.

The European Investment Fund (EIF)

The EIF was established in 1994 and its main shareholder is the EIB. The EIF supports, in particular, small and medium enterprises (SMEs) through direct loans, guarantees, private equity, venture capital and microfinance.[58]

[53] European Investment Bank (EIB) (2015).

[54] European Investment Bank (EIB) (2016).

[55] See also United Nations (UN) (2015), European Commission (EC) (2018b), European Investment Bank (EIB) (2015, 2016).

[56] For example, senior loans and guarantees, subordinated loans and guarantees, mezzanine finance and project-related derivatives.

[57] For example, the Mediterranean Hot Spots Investment Programme (MeHSIP).

[58] Maximum EUR 25,000 to micro-enterprises.

Moreover, it provides advising and sponsoring on behalf of third-party investors, including governments and private investors. The EIF operates in EU member states, potential EU member states as well as Liechtenstein and Norway. Besides supporting SMEs as main repositories of the EU economy and finance, the EIF is committed to sustainability through several initiatives. As an example, the Social Impact Accelerator (SIA) was Established in July 2015 by the EIF at the size of EUR 243 million and it provides equity to social enterprises, namely SMEs whose activity is self-sustainable and includes reinvesting growth to achieve a social impact (social inclusion, employment for marginalised groups and contribution to national growth).

6.6 EU and International Cooperation on Sustainable and Green Finance

6.6.1 Institutional Cooperation with the People's Republic of China

International cooperation on sustainable and green finance is fundamental for the development of the sector. Shared rules and official comparability devices among national and regional frameworks would allow for mobilising private capital consistently and with a global scope. In this respect, China has been providing a relevant contribution both as a large market for green securities and in terms of institutional commitment, especially with the EU. More precisely, the EIB and the CGFC[59] agreed that increasing volumes of green funds in China and EU convey the need of a harmonisation process (EIB and CGFC 2017, 2018), and hence cooperation should be fostered. However, their goal is not to define the "right" use of proceeds from green securities; rather, in an early phase the priority is given to the consistency among existing and upcoming standards, while a common language for green finance is indeed a long-term objective. In this regard, the first concrete step taken by the EIB and the CGFC was mapping and comparing the most used taxonomies for sustainable finance.[60]

Another impulse to the EU-China cooperation was given by the China's commitment to climate change since its presidency of the G20 summit in Hangzhou in 2016. On that occasion, China launched the G20 Green

[59] China Green Finance Committee, launched on April 2015.

[60] See also Chap. 2.

Finance Study Group (GFSG), which published the G20 Green Finance Synthesis Report in September 2016.[61] As a matter of fact, under China's presidency, green finance started to be a G20 key theme for the first time. In the Chinese national context, in 2016 the CGFC issued its Guidelines for Establishing the Green Financial System. They included as objectives: (i) vigorously develop green lending, (ii) enhance the role of the securities market in supporting green investment, (iii) launch green development funds and mobilise social capital through public and private partnerships (PPP), (iv) develop green insurance, (v) improve environmental rights trading market and develop related financing instruments, (vi) support local government initiatives to develop green finance, (vii) promote international cooperation in green finance and (viii) prevent financial risks and strengthen implementation. As concerns, in particular, the development of green bonds, the *Green Bond Endorsed Project Catalogue* (GBC)[62] of 2015 identifies sectors that are eligible for receiving green bonds' proceeds, the main categories being energy saving, pollution prevention and control, resource conservation and recycling, clean transportation, clean energy, ecological protection and climate change adaptation. Moreover, the issuance requires regulatory approval.[63]

China and Europe green finance systems still exhibit differences, even though efforts for a progressive harmonisation are today in progress.[64] However, once the EU taxonomy will be fully published, a transparent comparison among the two frameworks, a reinforced dialogue and con-

[61] Green Finance Study Group (GFSG) (2016). The report identified seven key actions: (i) Provide policy signals and frameworks, (ii) promote voluntary principles for green finance, (iii) expand learning networks, (iv) support local green bond markets, (v) promote cross-border green investments, (vi) encourage and facilitate knowledge sharing and (vii) improve impact measurement.

[62] Launched on December 2015 with the PBoC Announcement n. 39. Contributors were KPMG, the Global Reporting Initiative, the United Nations Environment Programme, and the Centre for Corporate Governance in Africa.

[63] In particular, during preparation, after having identified qualifying projects and disclosed the identification process, an external independent review is often employed. During issuance, the support investment banks and yearly reporting are recommended. As a priority action (likewise the EU policy), a green industry catalogue to better clarify definitions will also be released by 2019 as in the case of the EU taxonomy. Once available, the PBoC may credibly launch several initiatives, as including green bonds and highly-rated green loans as eligible collaterals for Central Bank borrowing, giving higher macro prudential scores to greener banks and lowering risk weights for green loans.

[64] At the country pair level, China and UK have been carrying harmonisation efforts as documented in the UK-China Green Finance Taskforce (2017).

vergence, and subsequent credible cross-issuances will be relevant and concrete contributions to further globalise green finance.

6.6.2 EU Financial Instruments in the International Context

European Fund for Sustainable Development (EFSD)

The EFSD was founded in September 2017 and includes a financial guarantee (of EUR 1.5 billion) and other blending instruments. It is one of the financial pillar of the EU's external investment plan, which since September 2016 supports the EU Neighbourhood[65] and sub-Saharan Africa. The EFSD provides technical assistance and interacts with governments, businesses and stakeholders to foster job creation, entrepreneurship, green and inclusive growth to foster gender equality, good governance and human rights and equitable access to resources. EFSD improves project sustainability, quality and innovation by investment grants (funding components or a percentage of the project), interest rate subsidies (reduction of the investment cost), technical assistance (ensuring high quality, efficiency and sustainability), risk capital (improving the risk/return profile to attract new funders) and guarantees (transferring pre-agreed risks to third parties with better absorption capacity). In July 2018, reported EFSD blending was of EUR 1.3 billion. In sub-Saharan Africa, blending intervention was of EUR 900 million,[66] leveraging EUR 5.6 billion. In the EU Neighbourhood, blending was of EUR 400 million,[67] leveraging EUR 5 billion.

Global Energy Efficiency and Renewable Energy (GEEREF)

Established in 2008 in Luxembourg, the GEEREF is a fund-of-funds investing in renewable energy and energy efficiency. Its policy is equally focused on enabling access to sustainable energy, climate and environment protection and achieving persistent returns. The GEEREF operates through private equity funds[68] focused on SMEs that qualify for the so-called Official Development Assistance. It is advised by the EIF, which provides expertise in climate-related activities and fund-of-funds management, and by the EIB.

[65] Algeria, Armenia, Azerbaijan, Belarus, Egypt, Georgia, Israel, Jordan, Lebanon, Libya, Moldova, Morocco, Palestine, Regional East, Regional South, Syria, Tunisia, and Ukraine.

[66] Thirty projects in transport (47.9%) energy (36.2%), private-sector financing (9%), agriculture (6.6%), and urban development (0.3%).

[67] Twenty-two projects in private sector (29%), water/sanitation (21%), environment (17%), energy (12%), and other minor projects (21%).

[68] EUR 50–200 million to emerging markets. GEEREF invested in 10 funds across Africa, Asia, Latin America and the Caribbean.

Green for Growth Fund (GGF)
The GGF was established in December 2009 and was funded with the contribution of European institutions, among others. The GGF is a highly specialised fund supporting energy efficiency and renewable energy in south-eastern Europe, European eastern neighbourhood, Middle-East and North-Africa. At the end of 2018, total available funding was of EUR 470.7 million, and the average investment volume of about EUR 30 million. The GGF refinances financial institutions to enhance participation in sustainable projects. The activities of GGF also include a technical assistance facility that provides capacity building and training, validation and monitoring of environmental performance, and strategic advice to fund managers. The GGF discloses its environmental and social policy in detail.

Green Climate Fund (GCF)
Established in 2010 by the 194 UNFCCC countries, the GCF is an EIB sustainable investing partner and supports developing countries' adaptation to climate change. The GCF recognises the heterogeneity of priorities across countries. In this respect, a representative of the country where a project is foreseen takes an active part in the whole decision-making process. The GCF operates through its partner organisations (the so-called Accredited Entities[69]) and pays particular attention to the Least Developed Countries (LDCs), Small Island Developing States (SIDS) and African States. The fund's investments can be in the form of grants, loans, equity or guarantees, and includes a technical assistance facility that falls under the following areas: capacity building, enhancement of the environment, capacity building for adaptation measures, technical analysis for optimising environmental performance and public awareness. As of January 2019, the GCF committed EUR 4.6 billion[70] across 93 projects.[71]

[69] As of February 2019, the GCF works with 75 "Accredited Entities".

[70] With a pledged amount of EUR 10.3 billion.

[71] Asia Pacific (40%), Africa (36%), Latin America and Caribbean (18%), Eastern Europe (6%). By size: micro (12%), small (29%), medium (42%), large (17%). By financial instrument: grants (47%), loans (42%), equity (9%) or guarantees (2%).

Annex 6.1 EC Workplan of the Initiatives Set Out by the Action Plan

1. Establishing an EU Classification System for Sustainability Activities		
Subject to the results of its impact assessment, Commission legislative proposal on the development of an EU taxonomy for climate change, environmentally and socially sustainable activities	L	Q2 2018
Report of the Commission technical expert group providing a taxonomy for climate change mitigation activities	NL	Q1 2019
Report of the Commission technical expert group providing a taxonomy for climate change adaptation and other environmental activities	NL	Q2 2019
2. Creating Standards and Labels for Green Financial Products		
Report of the Commission technical expert group on a standard for green bonds	NL	Q2 2019
Commission delegated act on the content of the prospectus for green bond issuances	L2	Q2 2019
Assessment of applying the EU ecolabel to financial products	NL	As of Q2 2018
3. Fostering Investment in Sustainable Projects		
Building on the ongoing efforts to reinforce advisory capacity, including for developing sustainable infrastructure projects, the Commission will take further measures that will improve the efficiency and impact of instruments aiming at sustainable investment support in the EU and in partner countries.		
4. Incorporating Sustainability When Providing Investment Advice		
Subject to the results of its impact assessment, Commission delegated acts (MiFID and IDD) on the suitability assessment	L2	Q2 2018
ESMA to include sustainability preferences as part of its guidelines on the suitability assessment	NL	Q4 2018
5. Developing Sustainability Benchmarks		
Commission delegated acts on the transparency of the methodology of benchmarks and on the features of the benchmarks	L2	Q2 2018
Subject to the results of its impact assessment, an initiative creating a designated category of benchmarks comprising low carbon issuers	L/NL	Q2 2018
Report of the Commission's technical expert group on the design and methodology of the low-carbon benchmark	NL	Q2 2019
6. Better Integrating Sustainability in Ratings and Research		
Commission services report on progress made on the actions involving credit rating agencies	NL	Q3 2019
ESMA to assess current practices in the credit rating market; ESMA to include ESG information in its guidelines on disclosure for credit rating agencies	NL	Q2 2019
Study on sustainability ratings and research	NL	Q2 2019

(continued)

(continued)

7. Clarifying Institutional Investors and Asset Managers' Duties		
Subject to the results of its impact assessment, Commission legislative proposal to clarify institutional investors' and asset managers' duties on sustainability and to increase the transparency of end investors, including transparency on their strategy and climate-related exposures	L	Q2 2018
8. Incorporating Sustainability in Prudential Requirements		
Work towards incorporating climate risks into institutions' risk management policies and on the potential calibration of banks' capital requirements in the capital requirement regulation and directive to take into account climate change-related risks while safeguarding financial stability and ensuring coherence with the EU taxonomy	Issue under discussion in the ongoing legislative procedure	2018–2019
The Commission will invite the European Insurance and Occupational Pensions Authority to assess the impact of prudential rules for insurance companies on sustainable investment	NL	Q3 2018
9. Strengthening Sustainability Disclosure and Accounting Rule-Making		
Publication of conclusions of the fitness check on public corporate reporting. This will inform any future legislative action by the Commission	NL	Q2 2019
Revision of the guidelines on non-financial information as regards climate-related information	NL	Q2 2019
Subject to the result of its impact assessment, proposal requiring asset managers and institutional investors to disclose how they consider sustainability factors in their investment decision-making process (as part of the proposal foreseen under action 7)	L	Q2 2018
Establishing a European Corporate Reporting Lab as part of EFRAG	NL	Q3 2018
Commission to systematically request EFRAG to assess in its endorsement advice the potential impact of new or revised IFRS standards on sustainable investments	NL	Q1 2018
Commission request to EFRAG to explore sound alternative accounting treatments to fair-value measurement for long-term investment portfolios of equity and equity-type instruments	NL	Q2 2018
Commission report on the impact of IFRS 9 on long-term investments	NL	Q4 2018
10. Fostering Sustainable Corporate Governance and Attenuating Short-Termism in Capital Markets		
Assessment of possible ways to promote corporate governance more conducive to sustainable finance	NL	Q2 2019
ESAs to collect evidence of undue short-term pressure from capital markets on corporations and consider further steps based on such evidence	NL	Q1 2019

Notes: *L* legislative proposal, *L2* level 2 measure, *NL* non-legislative measure

Source: European Commission (EC) (2018b)

References

Battiston, S., Mandel, A., Monasterolo, I., Schütze, F., & Visentin, G. (2017). A climate stress test of the financial system. *Nature Climate Change, 7*, 283–288.

Directive 2014/95/EU. (2014). Amending directive 2013/34/EU as regards disclosure of non-financial and diversity information by certain large undertakings and groups. *Official Journal of the European Union.*

Directive 2013/34/EU. (2013). On the annual financial statements, consolidated financial statements and related reports of certain types of undertakings. *Official Journal of the European Union.*

European Commission (EC). (2007). *An energy policy for Europe.* COM(2007) 1 final. Brussels.

European Commission (EC). (2010). *EUROPE 2020 – A strategy for smart, sustainable and inclusive growth.* COM(2010) 2020 final. Brussels.

European Commission (EC). (2011). *Energy roadmap 2050.* COM(2011) 885 final. Brussels.

European Commission (EC). (2014). *A policy framework for climate and energy in the period from 2020 to 2030.* COM(2014) 15 final. Brussels.

European Commission (EC). (2016a). *Capital markets union – Accelerating reform.* COM(2016) 601 final. Brussels.

European Commission (EC). (2016b). *Impact assessment – Accompanying the document proposal for a directive of the European Parliament and of the council – amending directive 2012/27/EU on energy efficiency.* SWD(2016) 405 final. Brussels.

European Commission (EC). (2017a). *Consumer financial services action plan: Better products, more choice.* COM(2017) 139 final. Brussels.

European Commission (EC). (2017b). Guidelines on non-financial reporting (methodology for reporting non-financial information) – 2017/C 215/01. *Official Journal of the European Union.*

European Commission (EC). (2018a). *A clean planet for all – A European strategic long-term vision for a prosperous, modern, competitive and climate neutral economy.* COM(2018) 773 final. Brussels.

European Commission (EC). (2018b). *Action plan: Financing sustainable growth.* COM(2018) 97 final. Brussels.

European Commission (EC). (2018c). *Proposal for a regulation on the establishment of a framework to facilitate sustainable investment*. COM(2018) 353 final. Brussels.

European Commission (EC). (2018d). *Proposal for a regulation on disclosures relating to sustainable investments and sustainability risks and amending directive (EU) 2016/2341.* COM(2018) 354 final. Brussels.

European Commission (EC). (2018e). *Proposal for a regulation amending regulation (EU) 2016/1011 on low carbon benchmarks and positive carbon impact benchmarks.* COM(2018) 355 final. Brussels.

European Commission (EC). (2018f). *Evaluation of the European fund for strategic investments, of the European investment advisory hub, and of the European investment project portal.* SWD(2018) 316 final. Brussels.

European Investment Bank (EIB). (2015). *EIB climate strategy – Mobilising finance for the transition to a low-carbon and climate-resilient economy.*

European Investment Bank (EIB). (2016). *The EIB group operational plan 2017–2019.*

European Investment Bank (EIB). (2017). *Climate action – List of eligible sectors and eligibility criteria.*

European Investment Bank (EIB) and Green Finance Committee of China Society for Finance and Banking (CGFC). (2017). *The need for a common language in green finance – Towards a standard-neutral taxonomy for the environmental use of proceeds – China's green bond endorsed project catalogue and the common principles for climate mitigation finance tracking mapped and compared.* White Paper, Phase I Report.

European Investment Bank (EIB) and Green Finance Committee of China Society for Finance and Banking (CGFC). (2018). *The need for a common language in Green Finance – Towards a standard-neutral Taxonomy for the environmental use of proceeds – Extending classification of objectives and activities for a shared market-based measurement of environmental impact and risk.* White Paper, Phase II Report.

European Union (EU). (2007). *Berlin declaration on the occasion of the 50th anniversary of the signature of the treaties of Rome (1957–2007).*

European Union (EU). (2015). Regulation (EU) 2015/1017 of the European Parliament and of the Council on the European fund for strategic investments, the European Investment Advisory Hub and the European investment project portal. *Official Journal of the European Union.*

European Union (EU). (2017). Regulation (EU) 2017/1129 on the prospectus to be published when securities are offered to the public or admitted to trading on a regulated market. *Official Journal of the European Union*

Financial Stability Board (FSB). (2017). *Recommendations of the task force on climate-related financial disclosures.* Final Report from the industry-led TCFD.

Fitch. (2017). *Green bonds – Fitch ratings and market overview.*

G20 Green Finance Study Group (GFSG). (2016). *G20 Green Finance Synthesis Report.*

High-Level Expert Group on Sustainable Finance (HLEG). (2017). *Financing a sustainable European economy.* Interim Report by the High-Level Expert Group on Sustainable Finance – Secretariat provided by the European Commission.

High-Level Expert Group on Sustainable Finance (HLEG). (2018). *Financing a sustainable European economy*. Final Report by the High-Level Expert Group on Sustainable Finance – Secretariat Provided by the European Commission.

Migliorelli, M., & Dessertine, P. (2018). Time for new financing instruments? A market-oriented framework to finance environmentally friendly practices in EU agriculture. *Journal of Sustainable Finance and Investment, 8*(1), 1–25.

Presidency Conclusions (PC) at the Lisbon European Council of March 23rd and 24th, 2000.

Technical Expert Group on Sustainable Finance (TEG). (2019a). *Report on climate-related disclosures*. Brussels.

Technical Expert Group on Sustainable Finance (TEG). (2019b). *Taxonomy. Technical report*.

The Lancet. (2017). *The lancet countdown on health and climate change: From 25 years of inaction to a global transformation for public health* (Vol. 391, No. 10120). https://www.thelancet.com/climate-and-health

UK-China Green Finance Taskforce. (2017). *Turning green momentum into actions*. Interim Report.

United Nations (UN). (2015). *Transforming our world: The 2030 agenda for sustainable development*, A/RES/70/1. New York.

United Nations Framework Convention on Climate Change (UNFCCC). (2015). *Paris agreement*. 21st Conference of the Parties (COP-21). Paris.

PART II

The (Long) Way Forward and New Opportunities

CHAPTER 7

From *Transaction-Based* to *Mainstream* Green Finance

Marco Migliorelli and Philippe Dessertine

7.1 Introduction

Green finance still has a long way to go before it can be considered a significant and stable component of the modern financial landscape. In this respect, the early success of the first green securities (in particular of *green bonds*), mainly

The authors would like to thank Rosita Fagiolo for her support to this chapter and the entire project of the book.

The contents included in this chapter do not necessarily reflect the official opinion of the European Commission. Responsibility for the information and views expressed lies entirely with the authors.

M. Migliorelli
IAE Université Paris 1 Panthéon-Sorbonne (Sorbonne Business School), Paris, France

European Commission, Brussels, Belgium

P. Dessertine (✉)
IAE Université Paris 1 Panthéon-Sorbonne (Sorbonne Business School), Paris, France
e-mail: philippe.dessertine@univ-paris1.fr

M. Migliorelli, P. Dessertine (eds.), *The Rise of Green Finance in Europe*, Palgrave Studies in Impact Finance,
https://doi.org/10.1007/978-3-030-22510-0_7

due to the strong market demand coming from institutional investors, risks to convey the idea that a smooth consolidation of the market will occur in the near future. Nevertheless, such a natural evolution is unlikely to happen. Indeed, it seems more reasonable to argue that, to take advantage of the positive momentum that green finance has been experiencing for the past years, market dynamics would need to be supplemented by policy and industry actions. To this extent, a number of initiatives have been launched by governments and international organisations worldwide, featured by different levels of ambition and perimeters. Yet, even in the most promising approaches, at least a few years will be needed before the envisioned measures eventually start producing the desired effects. In addition, some of the possible policy interventions that would still be necessary to effectively mainstream green finance require an in-depth preliminary assessment based on data and research which is currently only in a very early stage of development (e.g. as concerns the analysis of the relation linking environmental risks and economic and financial risks).

In such a context, the objective of this chapter is twofold. First, it aims at giving a systemic overview of the main challenges still ahead for green finance. Second, it discusses possible ways to deal with these challenges and identifies some key priority areas. Even if many of the issues analysed in this chapter concern the green finance market in general, a specific focus should be made on the situation in the European Union (EU). In fact, the European Commission (EC) has already identified a wide set of actions having as a specific object to further develop the whole sustainable finance sector (e.g. EC 2018)[1,2] and has started preparing some of the related legislative proposals.[3] This set of actions, which bring the EU at the forefront in terms of policy engagement, may represent a concrete basis for strengthening both green finance and the other components of the sustainable finance market. In some cases, once fully implemented, they may address some of the fundamental issues discussed in this chapter. For this reason, a reference to these initiatives is given whenever necessary.

[1] Sustainable finance can be broadly defined as the stocks and flows of financial resources and assets (across banking, investment and insurance industries) which is aligned with a large range of environmental, social and economic objectives and more generally with the delivery of the Sustainable Development Goals (SDGs) as developed in the context of the United Nation Development Programme (UNDP). In this respect, green finance should be considered a fundamental component of sustainable finance. For more details, see also Chap. 2 and UNEP (2016).

[2] These initiatives are described in Chap. 6.

[3] The information contained in this chapter are updated as of June 2019. The reader can verify the presence of updates on the European Commission's activities for the development of sustainable finance here: https://ec.europa.eu/info/business-economy-euro/banking-and-finance/sustainable-finance_en#overview

7.2 *Transaction-Based* and *Mainstream* Green Finance

The first main challenge for policy makers with regard to the future of green finance refers to the fair assessment of its long-term role within the overall financial system, and in particular as concerns its effective contribution in fostering an environmentally sustainable economy. In this respect, two alternative position models can be considered as a reference: *transaction-based* green finance and *mainstream* green finance.

7.2.1 Transaction-Based *Green Finance*

Transaction-based green finance may be defined as *the stocks and flows of financial assets that are directed to finance environmental activities and are labelled as "green" following an independent decision of the concerned market investors.*[4] It can be argued that particularly outside the perimeter of action of international financial organisations, green finance has grown so far mainly in line with this model, that is thanks to the emergence over the years of concrete incentives for market investors to issue or buy green-labelled securities. In this respect, three main incentives can be observed: reputational gains, corporate social responsibility acknowledgement (by existing and prospective clients) and the realisation of specific business opportunities. As concerns the entangled aspects of reputational gains and social corporate responsibility acknowledgement, abundant evidence exists today proving that companies may derive tangible benefits (such as better client loyalty or the possibility to enforce a market premium for their products or services) from being perceived as engaged in sustainable activities (e.g. Anselmsson et al. 2014), and that a generally positive relationship has existed at least in the last 20 years between corporate finance performances and environmental, social and governance (ESG) performances (e.g. Friede et al. 2015).[5] In addition, an even stronger commitment towards high standards of reputation and corporate social

[4] Here we refer to investors as the ensemble of potential issuers and potential buyers of green securities.

[5] In many studies, this relationship has been proven even after taking into account the typical possible endogeneity problem (i.e. the fact that the most financially successful companies may be the ones that decide to be involved in sustainable initiatives). See again Friede et al. (2015).

responsibility by embracing sustainability should be expected today by the growing number of organisations with formal or informal dual bottom-line strategies[6] (e.g. Migliorelli 2018). On the other hand, for some actors, green finance has emerged as an incremental or completely new source of revenues. This has been the case in particular for the investment funds industry (which could start offering green-labelled investment solutions to their clients), stock exchanges (that launched green listing options and related services) and green certification agencies (which could initiate new labelling and ancillary activities).[7] As a matter of fact, the presence of a concrete business opportunity has played a decisive role in encouraging these market actors to sustain the development of green finance in recent years.

Despite the mentioned incentives and even considering that *transaction-based* green finance can likely count today on a significant unexploited potential, it can be stated that such a model may face some limits when it comes to meeting the policy objective of mobilising a *sufficient* amount of financial resources to reach the most ambitious environmental targets.[8] In this respect, the most important drawback of a *transaction-based* model should be found in the inherent perspective of market investors and in the specific role of environmental risks in their decision-making processes. In fact, as of today, investors do not need to *fully* take into account the potential impacts of their investment decisions on the environment, since some of the related risks eventually materialise only over a long-time horizon and in many cases result in negative externalities for which they do not pay

[6] Organisations that have a dual bottom-line approach seek to extend the conventional bottom line, which measures economic and financial performance, by adding a second bottom line to measure their performance in terms of positive social impact. Ethical finance organisations and cooperative banks are examples of dual bottom-line organisations.

[7] As regards other possible incentives, little evidence still exists in particular concerning the presence of economic and financial benefits linked to the issuance of green securities (e.g. in the case of a lower cost of debit in case of the issuance of *green bonds*). Nevertheless, this situation may rapidly change, as a growing body of literature is emerging focussing on the impact of green labelling on the performances of financial securities. For more details, see Chap. 5.

[8] Possible early signs of a reduction in the growth of the green finance market are given by the green bond market in 2018, which experienced a reduction in the issuance in the first nine months of the year with the respect to the same period of 2017 (USD 108.3 bn, falling short of 2017 by about 1%). Source of data: Climate Bonds Initiative (2018).

for directly (e.g. the costs for the treatment of pollution-related diseases).[9,10] Hence, it should be argued that the market incentive for investors of *entirely* factoring-in environmental risks in their investment strategies (including the aspects of the selection of the projects to be undertaken and the way they are funded—e.g. via the issuance of green securities) is today limited. In point of fact, a structurally lower level of investments in environmental activities with respect to ideal environmental-neutrality standards is implied in any model based principally on market dynamics.

Similarly, other important limits of a *transaction-based* green finance model can be found in the cost of engagement for potential issuers and in the attitude of (some) financial market investors to consider green finance as a typical example of suboptimal investment choice. Due to screening, labelling, disclosure and control-related activities, issuing green securities conveys a not negligible administrative burden, which may represent a significant financial (and organisational) hurdle, in particular for smaller economic agents. In this respect, high transaction costs can explain why the issuance of green-labelled securities today remains restricted to large companies and international financial organisations. On the side of the demand, it can be observed that the attitude of some investors to consider green finance as a suboptimal investment choice is mainly due to the fact that only a relatively limited subset of the available investment universe is normally eligible as a concrete investing option (i.e. green securities). Following a traditional Markowitz approach to portfolio selection, these

[9] In this respect, the EU's emissions trading system (ETS) needs to be mentioned as a way to force companies to internalise at least part of these risks. Launched in 2005, the ETS is a tool that supports the effort to cut GHG emissions by at least 40% by 2030 compared to 1990 levels. It obliges more than 11,000 power plants and factories to hold (and pay) a permit for each tonne of CO_2 they emit, in order to provide a financial incentive to pollute less. Companies have to buy them through auctions and the price is affected by demand and supply (even though some of the permits are allocated for free, particularly in sectors at risk of having companies to move production to other parts of the world with laxer emission constraints). ETS encountered some technical issues linked to the unexpected fall of prices of permits after the crisis blasted in 2008, because of the drop-in demand, while the supply remained constant. In this respect, having a large surplus and low prices may effectively discourage companies from investing in green technology, thereby hampering the scheme's efficiency. In addition, this (useful) tool alone does not guarantee to reach the EU targets following the Paris Agreement.

[10] Furthermore, the methodologies for properly calculating environmental risks are still in an early phase of development. Even if a promising movement in research, supported by policy makers is emerging, little empirical results are available today.

Table 7.1 *Transaction-based* green finance: incentives for market investors and limits

Transaction-based *green finance*	
Incentives for market investors	*Limits*
Reputational gains Corporate social responsibility acknowledgement Realisation of business opportunities Proper management of environmental risks (for the part internalised)	Ineffectiveness in factoring in all the environmental risks in the investors' decision-making processes High administrative costs Restrictions of the investment spectrum in portfolio selection strategies

Source: Authors' elaboration

investors may consider the picking of large amounts of green securities as a strategy that may negatively affect the long-term risk-return profile of their portfolios.[11] Table 7.1 summarises the main incentives for market investors and limits of a *transaction-based* green finance model.

7.2.2 Mainstream *Green Finance*

Conversely, *mainstream* green finance may be defined as *the stocks and flows of financial assets that are directed to finance environmental activities in an amount sufficient to effectively contribute to reach an environmentally sustainable economy.* In this respect, *mainstream* green finance should be considered the target model for the most ambitious policies in terms of contribution of financial markets to environmental sustainability (see also Table 7.2). Nevertheless, it can be easily argued that no single *silver-bullet* exists in terms of technical solutions to mainstream green finance. Indeed, a combined set of actions should be put in place. These actions should allow at least three key conditions to be met:

- Environmental risks are properly included in the investors' decision-making processes.
- Market demand is effectively channelled towards green investments.
- *Additionality* is adequately encouraged by policy makers.

[11] Nevertheless, such a view is rapidly evolving. See Chap. 5 for a more detailed discussion of the financial performances of green securities.

Table 7.2 Comparison between *transaction-based* and *mainstream* green finance

	Transaction-based *green finance*	Mainstream *green finance*
Expected contribution to reach a fully environmentally sustainable economy	Limited	Relevant
Consolidation of the market	Induced by market incentives	Fostered by policy makers
Level of penetration of financial markets by labelled green securities	Relatively low	Medium-high
Additionality	Possible	Necessary
Cost of green finance for market investors	High and borne only if benefits are higher	Depends of policy requirements
Role of green finance in banking and insurance operations	Limited	Relevant

Source: Authors' elaboration

Today, one of the major obstacles to the proper inclusion of environmental risks in the decision-making processes of market investors is represented by the lack of reliable and usable information on the effective impact of these risks on economic or financial activities.[12] Such uncertainty first (and maybe foremost) limits the political and technical capacity of policy makers to shape appropriate adjustment mechanisms (e.g. a fair carbon tax for polluters). As a matter of fact, also in the field of sustainability, the effectiveness of the policy action is directly proportional to the reliability and the quantity of the evidence available to explain a specific phenomenon. Nonetheless, the lack of reliable and usable information has the potential to result in particularly serious consequences on those economic sectors that indirectly have to face environmental risks, as they are implicitly part of their business model.[13] This is especially the case for the

[12] This also contributes to the above-mentioned lack of incentives for market investors to fully internalise some of these risks.

[13] This issue has been also addressed by the Central Banks and Supervisors Network for Greening the Financial System (NGFS), a group of Central Banks and Supervisors willing, on a voluntary basis, to exchange experiences, share best practices, contribute to the development of environment and climate risk management in the financial sector, and to mobilise mainstream finance to support the transition towards a sustainable economy. The Network brings together Banco de España, Banco de México, Bank Al Maghrib, Bank of England, Bank of Finland, Bank Negara Malaysia (Central Bank of Malaysia), Banque Centrale du Luxembourg, Banque de France/Autorité de Contrôle Prudentiel et de Résolution (ACPR), Bundesanstalt für Finanzdienstleistungsaufsicht (BaFin), De Nederlandsche Bank, Deutsche

insurance and banking industries. For some observers, the failure to correctly price environmental risks by the various components of the financial industry may even represent in the longer term a threat to the stability of the financial system, as it can result in significant unexpected losses for financial intermediaries (e.g. EC 2018).[14,15] This reasoning is backed by the consideration that, even if it can be today stated that climate changes due to greenhouse gas (GHG) emissions ultimately result in a substantial increase in the frequency and magnitude of climate-related extreme weather events (e.g. IPCC 2018),[16] scarce information is still available on how *more frequent* and *more intense* these events may be. Hence, as extreme weather events may cause important reductions in the productivity of the economic assets in the regions affected, a scenario in which insurance companies will need to face unexpected higher levels of payments on the previously insured risks[17] and banks higher levels of impair-

Bundesbank, European Banking Authority, European Central Bank, Finansinspektionen (Swedish FSA), Japan FSA, Monetary Authority of Singapore, National Bank of Belgium, Oesterreichische National Bank, the People's Bank of China, the Reserve Bank of Australia, Reserve Bank of New Zealand.

[14] In addition, the lack of data on the impact of environmental risks may also generate situations in which financial intermediaries may decide to limit the provision of credit or insurance services to business or households potentially more concerned by these risks.

[15] The unexpected increase in the frequency and magnitude of climate-related extreme weather events is only one of the possible factors linked to climate change and potentially having an impact on the stability of the financial sector. In this respect, even policy actions can also bring some risks. This is the case when the physical assets used as collaterals for banking loans lose value resulting in non-compliance with new stricter environmental regulations (e.g. a building failing to respect new energy efficiency standards), or when the creditworthiness of banks' clients owning physical assets decreases following a change in national policy (e.g. the owner of a coal power plant to be closed before expected and hence representing as a *stranded asset*).

[16] The mentioned report (IPCC 2018) gives a detailed analysis of the consequences of an increase of the global temperatures of 1.5 °C with respect to pre-industrial levels. Examples of climate-related extreme weather events are droughts, floods and storms. Some early evidence is also available on specific economic impacts. Based on Lancet (2017), between 2000 and 2016, annual weather-related disasters worldwide rose by 46% and, between 2007 and 2016, economic losses from extreme weather worldwide rose by 86% (EUR 117.0 bn in 2016).

[17] Or new risks insured on the basis of data not taking into account climate-related extreme weather events.

ments on credits due to higher rates of insolvency of their clients[18] could eventually materialise.

On the contrary, the possibility of coherently (e.g. on the basis of reliable data) incorporating environmental outcomes in risk and pricing assessments would represent both an effective mechanism to mitigate the systemic risks and an instrument to stimulate *mainstream* green finance. As concerns the financial industry, it might result in a lower cost of financial services for environmentally conscious firms or projects and an increased burden for high-polluting ones.[19] Hence, it should be claimed that a thorough analysis of the relation linking environmental risks and economic and financial risks has to be one of the key areas of action for policy makers (and academicians) in the next few years. In this respect, and in the attempt to face (part) of the mentioned issues, from the EU perspective an action has been recently launched by the European Commission. The aim is to better integrate sustainability into risk management, in particular by exploring the possibilities given by incorporating environmental factors in existing rating systems and by better developing market research (EC 2018).[20] Nevertheless, such an action can be expected to eventually produce concrete results only in the mid-term.

[18] In more detail, climate-related extreme weather events can produce damages that can result in a reduction of the value of the assets held by financial institutions and of the collateral backing banks' loans, as well as in the impossibility of firms and households to repay what they have borrowed.

[19] Such an approach will avoid the alternative (but indeed possible) option through which banks and insurances companies would charge higher interest rates on credits and higher premiums on insured risks to clients irrespectively from their effective contribution to generate environmental risks. In point of fact, even though this option will probably reduce the systemic risks on the bank and insurance industries linked to environmental risks, it will result in an undue burden on non-polluting firms and households and would not represent an incentive for polluters to improve their environmental footprint. In addition, no reduction of environmental-related risks will eventually occur.

[20] Action 6 of the action plan for financing sustainable growth. In this respect, the Commission plans to (i) engage with all relevant stakeholders to explore the merits of amending the Credit Rating Agency Regulation to mandate credit rating agencies to explicitly integrate sustainability factors into their assessments in a proportionate way to preserve market access for smaller players, (ii) invite the European Securities and Markets Authority (ESMA) to assess current practices in the credit rating market, analysing the extent to which environmental, social and governance considerations are taken into account and to include environmental and social sustainability information in its guidelines on disclosure for credit rating agencies and consider additional guidelines or measures, where necessary and (iii) carry out a comprehensive study on sustainability ratings and research to analyse methodolo-

The second condition to foster ***mainstream*** green finance concerns the appropriate channelling of the existing and latent market demand. As a matter of fact, potential buyers (in particular if small economic agents) need reliable and easy-to-understand references in order to eventually direct their financial resources towards green securities. In the absence of such references, the cost of screening, analysis, selection and control of the effective use of proceeds necessary to ensure to effectively invest in instruments aiming at financing environmental activities would result in costs too high to bear. In this respect, a combined set of actions can be put in place. The identification of a unique definition of green activities (e.g. through a taxonomy), specific labelling criteria for green financial products,[21] disclosure requirements in public corporate reporting and the incorporation of the clients' preferences with respect to green finance in the counselling activities of financial intermediaries[22] are all factors that can jointly contribute to effectively steer market demand towards green securities. To this end, the financial industry has already endogenously developed some of these references (in particular, private green labels for bonds and, as concerns large corporations, certain voluntary disclosure elements on sustainable activities), which have indeed contributed to stimulate the development of the market. In addition, and even more importantly, a coherent set of actions, covering all the above-mentioned elements, has been already foreseen by the European Commission and will be progressively developed in the years to come (EC 2018).[23]

gies and explore aspects like the market structure of sustainability ratings and market research services, the depth and breadth of sustainability research assessments and scoring, and the independence of those research/scoring providers. See also Chap. 6.

[21] Considering the emergence of green labels in the financial markets, it is worth mentioning that mainstreaming green finance only partially implies labelling securities as green securities. Indeed, labels may represent useful tools for market players to easily screen the possible investment options and hence steer the market demand. Nevertheless, they imply an increased cost for issuers and their use should be constantly monitored in order to avoid opportunistic behaviours (as, but not limited to, the practice of ***greenwashing***). As a matter of fact, in the policy perspective green labels should be henceforth considered as instruments more than final objectives.

[22] The counselling activity of financial intermediaries is regulated by the Market in Financial Instruments Directives and Market in Financial Instruments Regulation (MiFID I, MiFID II and MIFIR), in particular *vis-à-vis* retail clients, for which an analysis of suitability and appropriateness of the investment is mandatory.

[23] With respect to the European Commission action plan for financing sustainable growth (EC 2018), this is the case of the establishment of an EU classification system for sustainable activities (action 1), the creation of standards and labels for green financial products (action 2),

Finally, in order to ensure that green finance will effectively contribute to reach a sustainable economy, it is important that ***additionality*** is encouraged when needed. Here, with the term ***additionality*** it is intended a policy-triggered incremental stream of financial resources directed towards a specific objective (in this case, the financing of environmental activities). In this respect, it can be argued that the possible and foreseen actions aiming at properly including environmental risks in the investors' decision-making processes and at steering the market demand towards green investments would hardly be completely effective in providing the ***sufficient*** amount of resources needed for financing the environmental transition of the economy. In fact, even in the jurisdictions in which they will be implemented, a relatively long transition period will be needed before the foreseen actions start delivering concrete results. More likely, a certain level of ineffectiveness should be expected.[24] In some cases, such ineffectiveness may add to sector-specific market failures already affecting the financing capabilities of the economic agents that are supposed to carry over the environmental investments.[25] These elements may indeed justify a policy intervention directed to stimulate an adequate level of funding for financing environmental activities. To do that, it might be also useful to introduce in the policy-making process the concept of ***environmental market***

the incorporation of sustainability when providing financial advice (action 4), the strengthening sustainability disclosure and accounting rule-making (action 9). Following the plan, in May 2018 the European Commission adopted a package of implementing measures including: a proposal for a regulation on the establishment of a framework to facilitate sustainable investment (this regulation establishes the conditions and the framework to gradually create a unified classification system—or taxonomy—on what can be considered an environmentally sustainable economic activity); a proposal for a regulation on disclosures relating to sustainable investments and sustainability risks and amending Directive (EU)2016/2341 (this regulation aims at introducing disclosure obligations on how institutional investors and asset managers integrate environmental, social and governance factors in their risk processes); a proposal for a regulation amending the benchmark regulation (to create a new category of benchmarks comprising low-carbon and positive carbon impact benchmarks, which should provide investors with better information on the carbon footprint of their investments). See also Chap. 6.

[24] It can also be argued that some of the environmental objectives may be left uncovered in the case investors factor-in only the environmental risks having some form of economic or financial impact. In fact, some of the environmental risks (e.g. the loss of the biodiversity) may have only a very weak relation with economic and financial risks. Similarly, factoring-in environmental risks to the extent to which they have economic or financial consequences may still produce a degradation of the environment of a higher level than societally acceptable.

[25] See, for example, Migliorelli and Dessertine (2018).

failure, highlighting situations in which market-based dynamics are not sufficient to generate an adequate stream of resources towards environmental investment. In operational terms, fostering ***additionality*** may be done by implementing dedicated financing facilities, backed by lending, guarantees or equity participation (or a combination of the three).[26,27]

7.3 Fostering a Green Banking Market by Encouraging the Use of *Green Loans*

The effective involvement of the banking sector in the development of green finance is of utmost importance. The relevance of banks is given in particular by the high level of funds intermediated and by their unique capacity to reach small and medium enterprises (SMEs) and households, which are often pivotal actors for the success of sustainable policy actions (and usually do not have other sources of financing than bank lending).[28] The involvement of the banking sector in promoting green finance may even be considered essential in Europe. In fact, the incidence of banks in the continent is substantially higher than in other economic systems, for example the United States, where the share of intermediated funds is much lower in terms of GDP.[29,30]

[26] See for example the case of the agriculture discussed in Chap. 8.

[27] Nevertheless, such an intervention should be justified by a detailed assessment in order to verify the effective presence of market failures or suboptimal investment situations. In the EU this practice is already adopted by policy makers in several programmes, through the so-called *ex-ante assessments* (see, e.g. the *Methodological handbook for implementing an ex-ante assessment of agriculture financial instruments under the EAFRD*). In this respect, policy-driven financing in the absence of market failures or suboptimal investment situations may result to go against state's aid legislation.

[28] See for example Beck et al. (2009). The attitude of banks in financing SMEs and households is particularly strong in stakeholders-value banks, that is, cooperative and savings banks. See, for example, Cornée et al. (2018) for a recent overview of the functioning of these kinds of banks.

[29] The total assets of the European banking sector peaks up to 350% of aggregate GDP, while the same metrics is about 77% in the United States. Source: EBF (2012).

[30] The importance of banking in the context of sustainable finance has been recognised also by the United Nations Environment Programme (UNEP), which in 2018 identified six *Principles for Responsible Banking*, referring to: alignment (to the Sustainable Development Goals, the Paris Climate Agreement and relevant national and regional frameworks), impact, clients and customers, stakeholders, governance and target setting, transparency and accountability.

The three key conditions needed to mainstream green finance (environmental risks are properly included in the investors' decision-making processes, market demand is effectively channelled towards green investments and ***additionality*** is adequately encouraged) find a common ground in banking operations. In this respect, an argument can be made according to the idea that one of the main factors able to mainstream green finance within the banking sector refers to a coherent recognition and treatment of the loans issued to finance environmental activities. These loans could be eventually defined as ***green loans***.[31] The possibility to easily detect such type of assets in the banks' balance sheets may bring two main advantages. On the one hand, it would contribute to the proper management of environmental risks at the banking level. On the other hand, the identified ***green loans*** could be used as underlying assets for the arrangement of structured finance instruments suitable for a larger public of investors, even outside the banking sector.

With regard to the recognition of ***green loans*** as contributors to the management of environmental risks at the bank level, it can be expected that such a recognition may represent a key enabler for the sought incorporation of environmental factors in risk and then in pricing of the loans. By providing a structural segregation of the different types of loans by the use of proceeds, the recognition of ***green loans*** would allow banks to progressively set a different cost of funding for environmentally friendly investments. Nevertheless, managing ***green loans*** would require banks to perform additional screening, labelling, disclosure and control-related activities. As of today, for banks the cost of such a transformation would likely be much higher than the potential benefits. As a result, it can be expected that banks will continue to have little incentive to put in place

[31] The very first definitions of ***green loans*** have already emerged in the market, even though they are scarcely or not used in the banking sector. In particular, in March 2018 the Loan Market Association (LMA) published the first Green Loans Principle (GLPs), mainly mimicking the Green Bond Principles (GBPs) initiative. In the framework established by the LMA, green loans are defined as ***any type of loan instrument made available exclusively to finance or re-finance, in whole or in part, new and/or existing eligible green projects***. To be recognised as such, the concerned organisations have also to align with the four core management components of use of proceeds, process for project evaluation and selection, management of proceeds and reporting (LMA 2018). In addition, it seems reasonable to argue that the same taxonomy that could be developed for listing sustainable activities should also be used for the identification of ***green loans***.

the organisational structures and processes necessary to manage their lending activity according to any definition of *green loans* without a specific policy intervention. In this respect, at the EU level the European Commission has launched an action aiming at assessing the possibility to define a *supporting factor* as concerns capital requirements of banks, insurance companies and pension funds for the assets directed to finance sustainable activities (EC 2018), which could represent a first concrete regulatory incentive.[32,33] Nevertheless, concrete results can be expected only in the mid-term.

As concerns the second main advantage of recognising *green loans* at the bank level, that is, the possibility to build structured finance instruments backed by these loans, it can be argued that a wide range of technical options today exist. As included in the wider category of asset-backed securities (ABS), such instruments would mainly follow the principle of securitisation. A typical securitisation process allows assets that are normally illiquid (as loans) to be pooled and transferred to an ad hoc vehicle (so-called special purpose vehicle, SPV), which issues tradable securities backed by those assets. These securities are usually issued in tranches on the basis of different level of seniority (to appeal to a wider range of investors in terms of risk-return) and can be

[32] Action 8 of the action plan for financing sustainable growth. In this respect, the Commission plans to (i) explore the feasibility of the inclusion of risks associated with climate and other environmental factors in institutions' risk management policies and the potential calibration of capital requirements of banks as part of the Capital Regulation and Directive and (ii) invite the European Insurance and Occupational Pensions Authority (EIOPA) to provide an opinion on the impact of prudential rules for insurance companies on sustainable investments. The mentioned supporting factor will be eventually phased-in along with the establishment of an EU classification system for sustainable activities (action 1 of the plan). For more details, see Chap. 6.

[33] This action adds to the actions linked to the definition of a taxonomy for sustainable activities and specific labelling criteria for green financial products, already discussed and not focused on banks, insurance companies and pension funds. Finally, it would be indeed necessary that the classification of *green loans* at banks' level results from a process featured by low operational costs and high reliability. In this respect, it seems likely that the banking industry would be required to put in place specific operational processes and organisational structures (within the policy framework that will progressively consolidate) aimed at streamlining the concerned banking operations.

exchanged in the secondary markets, in particular in the bond market.[34] In addition, an external credit enhancer can be used to mitigate the risk for the final investors and create highly rated financial instruments.[35] By means of asset securitisation, the originating entity is cashed out, while the securities issued by the SPV are repaid through the cash flow of the underlying loans. The concrete benefits of fostering ABS having *green loans* as underlying assets should be expected in higher levels of capital flows directed to environmental investments and in lower financing costs for initial borrowers.[36] In fact, capital raised through the sale of asset-backed securities by the loan originators can be used to create a new portfolio of loans (e.g. IMF 2015) and labelling the ABS as *green* may enable issuers to further take advantage from the increasing demand for securities with environmental benefits.[37]

[34] Usually, at least three tranches are issued (senior, mezzanine and equity tranches) with a hierarchical right on the SPV repayments.

[35] In this regard, excess spread (the practice of issuing notes with an overall yield lower than that of the underlying assets) and overcollateralisation (the practice of issuing an amount of notes lower than the available underlying assets) are also used as sources of internal credit enhancement and to cover transaction costs linked to the securitisation operation.

[36] See also Duffie (2008) on the use of securitisation mechanisms backed by SMEs lending and short-term commercial papers, Caballero and Krishnamurthy (2009) on the creation of securities classes able to appeal to a large base of investors with different risk appetites and Cardone-Riportella et al. (2010) and Farruggio and Uhde (2015) on the benefits sought through securitisation as concerns the provision of new liquidity for financial intermediaries.

[37] Despite these advantages, existing literature highlights certain risks linked to securitisation practices. To this extent, a number of works have analysed the effects of the information asymmetries and the moral hazard that may feature the relationship between originator and final investors. In fact, banks and other financial institutions may tend to accept reducing their credit standards and transfer the risk to the market. In this respect, evidence has been documented, in particular for the subprime mortgages in the United States, which has been accused of triggering the financial crisis in 2007. Based on these studies, the absence of *skin in the game* has been the basis of a misalignment in the incentives between originators and final investors. This phenomenon has eventually caused a sensitive reduction of the quality of the underlying assets (e.g. Keys et al. 2009; Mian and Sufi 2009). For this reason, all recent regulation proposals on securitisation have included risk retention clauses concerning the originator. In more detail, the provision of maintaining a minimum nominal value of the first-loss tranche or of each of the tranches sold or transferred to investors is constantly proposed to limit opportunistic behaviours (BCBS 2014; IMF 2015; EC 2015).

7.4 Other Possible Facilitators for Mainstreaming Green Finance

7.4.1 Development of Green Products and Investment Options

The expansion of green finance is also dependent on the presence of an adequate supply of green-labelled securities and investment options appealing to a wide range of market investors. In this regard, two types of "traditional" financing arrangements might play an innovative role in the years to come. The first refers to the several aggregation possibilities of stand-alone financial products. The second is related to the development of public–private partnerships (PPP).

Aggregation can represent an important instrument to unlock additional capital flows, in particular from smaller-scale sustainable investments, most of which would otherwise be illiquid. In order to support the development of green finance, an ABS can either have green underlying collaterals (ABS eventually having *green loans* as underlying assets, as discussed in the previous section, are an example of this category), or it can have non-green collaterals but use the proceeds to invest in environmental projects. In point of fact, it should be recognised that a lot more flexibility would be given by ABS with non-green collaterals, along with lower screening cost in the absence of a widespread labelling of securities (as it is mainly the case today, in particular outside the *green bonds* market). Nevertheless, a specific regulation or market standards may be needed in order to avoid that non-homogenous forms of ABS emerge in the green finance sector, potentially harming the credibility of the nascent market.

As regards the development of specific public–private partnerships (and following the idea that market forces may not be systematically able to trigger an adequate level of investment towards environmental activities and hence *additionality* should be fostered), it can be argued that the role of the policy makers should be at least twofold. On the one hand, policy makers should identify the policy areas suffering from significant underfunding, in terms of both sectors and geography. On the other hand, they should encourage the participation of private investors to the financing of specific investments at the highest extent possible, in order to maximise the impact of the public spending whenever foreseen. In this regard and depending on the features of the project to be financed, a number of solutions have already been developed by governments and international financial organisations[38] that can be further applied to the specific case of the financing of environmental activities. A first one is represented by public organisations absorbing part of the risk (including by taking the first loss) on investments financed through dedicated lending, guarantees or equity facilities. In this way, private sector participation is encouraged thanks to the presence of a concrete risk cushion shielding the investment.[39] A second solution consists in the provision by public organisations of counselling and issuance-related services to private investors. In this respect, public organisations may play an important role in facilitating the structuring of a specific operation by effectively leveraging their centres of expertise. Advising on the basis of knowledge of market context, relevant legislation and technical requirements (including state-of-the-art governance and due-diligence standards) can often represent a key supporting factor to phase-in otherwise full-fledged private initiatives. Finally, public organisations may act as facilitators by playing a simple but pivotal endorsement role. The formal endorsement of public powers of a private investment may be beneficial for the success of an operation, thanks to the reputational benefits for private investors to be engaged in projects with high political visibility.

7.4.2 A Pipeline of Environmental Projects

Meeting the climate objectives would require the delivery of many new environmental projects across a range of technologies and in almost all sectors of the economy, implying the generation of a relevant additional flow of investments.[40] To do that, the availability of a robust pipeline of environmental projects ready to be financed should be considered as another facilitating factor to foster the environmental transition (and indirectly encouraging the development of green finance). This pipeline

[38] See also Arezki et al. (2017).

[39] Nevertheless, and depending of the specific structuring of the operation, provisions to avoid possible opportunistic behaviour of private investors and to adequately price the effective risk undertaken by the different parts have to be foreseen. This is valid in particular in the case governmental entities covering the project's first loss.

[40] The latest global estimates of investment needs may differ, but they all point to a financing gap of trillions of dollars per year until at least the year 2030. Only as concerns investments in green, low-carbon and climate-resilient infrastructures, investment volumes can be estimated to range from USD 3.4 trillion to USD 4.4 trillion globally (source: OECD 2018).

should be composed of a set of bankable projects for which financing arrangements have not yet been set up by the promoters or projects that would require to be financed or co-financed with public spending. In this regard, a typical obstacle to the swift creation within the private sector of a robust pipeline of suitable investments is that investors cannot always easily identify and source all the possible investment opportunities present in the market. In many cases, due to the lack of detailed investment plans and limited (or no) integration of these plans into a more general policy context, it may not be clear for investors which investments are needed, where they are needed, when they should be launched and how to finance them. This problem is particularly relevant for long-term projects, which are required to accurately fit within defined strategic priorities and are in many cases essential for reaching the climate objectives (as for example in the case of energy-related infrastructures).

Public actors may strongly influence the development of project pipelines. Nevertheless, there is no one single method to promote and develop such pipelines, as planning efforts vary greatly in scope and the scale of the investments may depend on specific country or regional contexts. In this respect, recent literature (e.g. OECD 2018) suggests a number of principles that, if applied, can facilitate the development of effective project pipelines. These principles include having a clear leadership to centralise coordination and manage the process (e.g. an investment hub at the government level), defining a strong enabling environment (i.e. a set of policies aligned across the economy, including financial support where necessary), assuring transparency of the process (including the identification of eligibility criteria to access public support and the prioritisation mechanisms to fast track highly valuable key initiatives) and establishing mechanisms to assure the dynamic adaptability of the pipeline (in order to maintain the pertinence of the investments even in changing external conditions and to avoid path dependency or lock-in effects). From the EU perspective, many of these principles have already been adopted in the context of the European Fund for Strategic Investments (EFSI), which nonetheless covers a much larger scope than environmental or sustainable initiatives.[41]

[41] In the EU, organisational structures have been created with the aim, inter alia, to foster a pipeline of projects to achieve the goals of the European fund for strategic investments (EFSI). These structures are the *European investment advisory hub* and the *European investment project portal*. The main purpose of the advisory hub is to provide advice to investors,

7.4.3 Green Finance Ratings

As of today, even in a context in which a growing number of green standards is being developed by the financial industry, market investors may still hardly answer the question of *how green* is a specific security that is exchanged in the market. In fact, even when securities are labelled as such (as in the case of *green bonds*), scarce easy-to-access information is available for investors as concerns the level of alignment to ideal best practices. In this respect, a more detailed information on the "shade of green" featuring a specific issuance may further contribute to build confidence in the green finance market, in particular once its first phase of sustained growth is concluded. In this respect, it can be expected that this confidence would mainly result from the appreciation by market investors of a thorough assessment performed by specialised bodies or agencies (whose trustworthiness hence represents an essential asset in the process).

Limited to the *green bonds* market, the financial industry has recently started developing the very first systems of green ratings. Unsurprisingly, the lead has been taken by existing certification and rating agencies, able to leverage both their technical expertise and their market positioning of incumbents in the sector.[42] In these systems, the assessment is related to specific *green bond* issuances and has as an objective the analysis of the features of the underling projects to be financed. In more detail, the level of compliance with the main categories of the use of proceeds, processes for project evaluation and selection, management of proceeds and reporting are typically scrutinised. Nevertheless, as of today, a very limited number of *green bond* issuances have been assigned with a green rating of this type and such a practice should still be considered to be in a very early stage of development. Hence, uncertainties in its long-term potential and effective market appeal remain.[43]

project promoters and public managing authorities on project identification, development and preparation. The advisory hub provides its services at both the EU and local level. The project portal's purpose is to provide information for potential investors about projects that require investors, to increase their visibility and thus contribute to the effectiveness of the European fund for strategic investments. The portal displays projects for information purposes only: the projects featured on the portal have not been pre-selected for financing from the European fund for strategic investments or EU programmes.

[42] CICERO's *Second Opinions*, S&P's *Global Ratings Green Evaluation* and Moody's *Green Bonds Assessment* have emerged as some of the main reference initiatives. CICERO uses four classes: dark green, medium green, light green and brown. S&P foresees five classes: GB1 (excellent), GB2 (very good), GB3 (good), GB4 (fair), GB5 (poor). Moody's considers four main classes (E1, E2, E3, E4) and an overall score out of 100.

[43] In this respect, it is worth mentioning that, as for ratings on debt issuances, the request to have a rating comes from the issuer.

7.4.4 *Academic Research on Green Finance*

Last, but not the least, the development of green finance would greatly benefit from a wider range of academic researches to be carried out in the different areas of attention for both the industry and policy makers. As a matter of fact, academic research on green finance may still be considered as a (promising) niche. Nevertheless, a thorough understanding of the specific features of this emerging sector would increase the possibilities of its widening within the overall financial market. In this respect, green finance has indeed the potential to consolidate as a completely new asset class already in the mid-term. Research questions such as "do green securities have a different risk-return profile vis-à-vis otherwise equivalent securities?", "do market investors effectively price a green factor embedded in market securities?", "which are the specific determinants of the risk and liquidity of a green security?" will need to be answered by academics in the next few years. Similarly, and as concerns in particular the management of systemic risks, questioning and providing evidence on "how environmental risks relate to financial risks?", "which are the channels of transmission of environmental risks to the financial sector?", "does sustainable finance effectively contribute to reach the sustainable development goals?" would have an increasingly relevant policy and societal impact. Hence, a combined effort from universities and policy makers should be encouraged (and specific initiatives put in place) in order to rapidly make academic research an essential contributor to an unbiased debate on the potentialities and limits of green finance.

References

Anselmsson, J., Bondesson, N. V., & Johansson, U. (2014). Brand image and customers' willingness to pay a price premium for food brands. *Journal of Product and Brand Management, 23*(2), 90–102.

Arezki, R., Bolton, P., Peters, S., Samama, F., & Stiglitz, J. (2017). Financing infrastructure. *Economic Policy, 32*(90), 221–261.

Basel Committee on Banking Supervision (BCBS). (2014, December). *Revisions to the Basel securitisation framework.* Basel.

Beck, T., Demirgüç-Kunt, A., & Martínez Pería, M.S. (2009). *Banking SME around the world: Lending practices, business models, drivers and obstacles.* World Bank Policy Research Working Paper, 4785.

Caballero, R. J., & Krishnamurthy, A. (2009). Global imbalances and financial fragility. *American Economic Review, 99*(2), 584–588.

Cardone-Riportella, C., Samaniego-Medina, R., & Trujillo-Ponce, A. (2010). What drives bank securitisation? The Spanish experience. *Journal of Banking and Finance, 34*, 2639–2651.

Climate Bonds Initiative. (2018, October). *Green bonds market summary*, Q3 2018. London.

Cornée, S., Fattobene, L., & Migliorelli, M. (2018). An overview of cooperative banking in Europe. In M. Migliorelli (Ed.), *New cooperative banking in Europe. Strategies for adapting the business model post-crisis*. London: Palgrave Macmillan.

Duffie, D. (2008). *Innovations in credit risk transfer: Implications for financial stability*. BIS Working Paper, 255.

EBF. (2012). *EU's banking sector: The world's largest banking system. Facts and figures 2011/2012*. Brussels.

European Commission (EC). (2015, September). *Proposal for a regulation of the European Parliament and of the council laying down common rules on securitisation and creating a European framework for simple, transparent and standardised securitization*. Brussels.

European Commission (EC). (2018). *Action plan: Financing sustainable growth*, COM(2018) 97 final. Brussels.

Farruggio, C., & Uhde, A. (2015). Determinants of loan securitization in European banking. *Journal of Banking and Finance, 56*, 12–27.

Friede, G., Busch, T., & Bassen, A. (2015). ESG and financial performance: Aggregated evidence from more than 2000 empirical studies. *Journal of Sustainable Finance and Investment, 5*(4), 210–233.

Intergovernmental Panel on Climate Change (IPCC). (2018). *Special report. Global warning of 1.5 °C*. Geneva.

International Monetary Fund (IMF). (2015, January). *Securitization: The road ahead*. IMF Staff Discussion Note.

Keys, B. J., Mukherjee, T., Seru, A., & Vig, V. (2009). Financial regulation and securitization: Evidence from subprime mortgage loans. *Journal of Monetary Economics, 56*, 700–720.

Lancet. (2017, October). *The Lancet Countdown on health and climate change: From 25 years of inaction to a global transformation for public health*. London.

Loan Market Association (LMA). (2018). *Green loan principles*. London: United Kingdom.

Mian, A., & Sufi, A. (2009). The consequences of mortgage credit expansion: Evidence from the 2007 mortgage default crisis. *Quarterly Journal of Economics, 124*(4), 1449–1496.

Migliorelli, M. (Ed.). (2018). *New cooperative banking in Europe. Strategies for adapting the business model post-crisis*. London: Palgrave Macmillan.

Migliorelli, M., & Dessertine, P. (2018). Time for new financing instruments? A market-oriented framework to finance environmentally friendly practices in EU agriculture. *Journal of Sustainable Finance and Investment, 8*(1), 1–25.
Organisation for Economic Cooperation and Development (OECD). (2018, November). *Developing robust project pipelines for low-carbon infrastructure*. Paris.
United Nations Environment Programme (UNEP). (2016. September). *Definitions and concepts. Background note*. Inquiry Working Paper 16/13.

CHAPTER 8

The Development of Green Finance in EU Agriculture: Main Obstacles and Possible Ways Forward

Marco Migliorelli

8.1 Introduction[1]

Agriculture[2] is responsible for about one-fourth of the total greenhouse gas emissions worldwide.[3] In this respect, it should be considered one of the most polluting sectors in the economy (Fig. 8.1).[4] Nevertheless, only

The contents included in this chapter do not necessarily reflect the official opinion of the European Commission. Responsibility for the information and views expressed lies entirely with the author.

[1] This chapter partially draws from Migliorelli and Dessertine (2018).

[2] In this chapter, the term *agriculture* also includes forestry and other land uses.

[3] Agriculture releases large quantities of carbon dioxide (through the burning of biomass mainly in areas of deforestation and grassland), is responsible for up to half of all methane emissions (livestock alone account for about a quarter of such emissions, while irrigated rice farming accounts for about a fifth of total emissions) and is a key source of nitrous oxide (generated by natural processes, but boosted by leaching, volatilisation and runoff of nitrogen fertilisers, and by the breakdown of crop residues and animal wastes). Source: http://www.fao.org/docrep/004/y3557e/y3557e11.htm

[4] In addition, crop and livestock production are the main source of water pollution by nitrates, phosphates and pesticides.

M. Migliorelli, P. Dessertine (eds.), *The Rise of Green Finance in Europe*, Palgrave Studies in Impact Finance,
https://doi.org/10.1007/978-3-030-22510-0_8

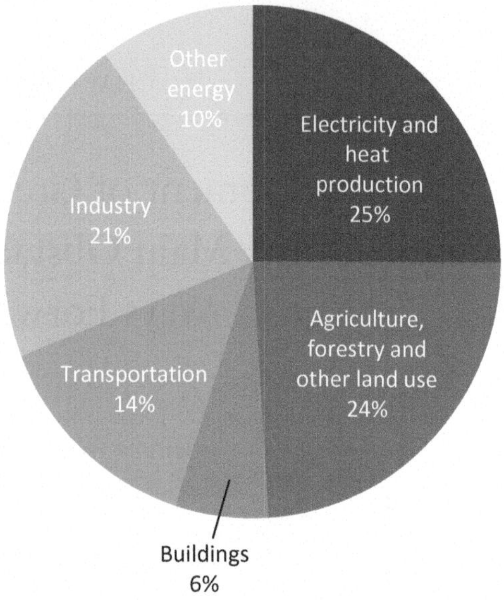

Fig. 8.1 Global greenhouse gas emissions by economic sector (2010). (Source: Author's elaboration on data IPCC (2014))

a negligible fraction of green securities is currently directed to finance agriculture (e.g. Fig. 8.2). This lack of green-labelled investments *vis-à-vis* perceivable needs may indeed contribute to delay the development of sustainable agriculture and the bio-economic transition.[5] In such a context, it is of utmost interest to analyse what are the main obstacles that still hamper the expansion of green finance in the agricultural sector today and to identify some actionable solutions to reverse the observed tendency. This chapter focuses on this challenge in the specific case of the European Union (EU). The choice to limit the investigation to the EU is principally due to the factual observation that the policy framework usually plays a

[5] As a matter of fact, the debate on how to foster sustainable agriculture and the bio-economic transition embraces many other disciplines. Aspects regarding effective regulation, technological improvements, scientific research and evolution in the lifestyle (in particular as concerns the types of proteins to be consumed, animal or vegetal) will also determine the feasibility and the speed of the changeover.

M. Migliorelli (✉)
IAE Université Paris 1 Panthéon-Sorbonne (Sorbonne Business School), Paris, France

European Commission, Brussels, Belgium

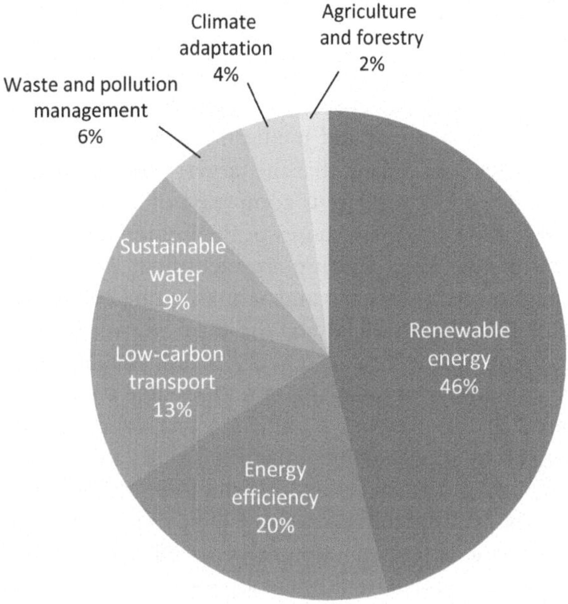

Fig. 8.2 Green bonds use of proceeds raised by sector (2015). (Source: Author's elaboration on data OECD (2017))

significant role in the way agriculture is financed. Worldwide, policy measures in agriculture include specific financing schemes (e.g. in the form of direct payments or price support) that strongly influence the behaviour of the affected economic agents, that is, agricultural producers (farmers). Hence, no thorough analysis on agricultural financing is usually possible without taking into account the specific characteristics and implications of the applicable public policies. In this respect, the actual system of support to agriculture in the EU is mainly axed on the Common Agricultural Policy (CAP) and, to a lesser extent, on a series of financing instruments managed by the European Investment Bank (EIB). EU agricultural producers tend to be highly dependent on direct public spending: the average share of direct payments to the agricultural factor income is about 28%, and taking into account all subsidies, total public support in agricultural factor income reaches on average 40%.[6]

[6] Data for 2010–2013. Agricultural factor income represents income generated by farming which is used to remunerate borrowed/rented factors of production (capital, wages and land rents) and own production factors (own labour, capital and land). Source: EC (2015b).

In order to perform the analysis, the reminder of this chapter is structured as follows. First (in Sect. 8.2), a review of the literature is carried out focusing on the green farming financing[7] in the EU. To this extent, the determinants of the farmers' choice to undertake green farming practices and the key features of small and medium enterprise (SME) lending are discussed, the latter considering that the largest portion of the agricultural producers in many EU countries is composed of small economic agents. Second (in Sect. 8.3), the main causes of the limited use of green-labelled financing instruments in agriculture are identified. To do that, we adapt the approach followed by Ashcraft and Schuermann (2008) and Migliorelli and Dessertine (2018), who analysed the main market failures existing in different financing chains linking borrowers and financial markets. Then (in Sect. 8.4), a number of financing structures able to overcome some of issues observed in the previous section are proposed. In particular, the possibilities given by securitisation, guarantee funds and funds of funds are discussed, including the necessary adaptations needed to tackle the specificities of the EU agricultural sector.[8] Hence (in Sect. 8.5), we discuss the possible intermediation function of cooperatives. In this respect, the role of both financial and agricultural cooperatives is evaluated. Finally (in Sect. 8.6), some concluding remarks are stated.

8.2 Literature Review

Existing research provides the general background for the investigation. Two main strands of literature are of particular relevance. On the one hand, the works related to the analysis of the determinants of the farmer's choice to undertake green farming practices, in particular as regards the aspects of the profitability of the investment and of the impact of the policy schemes. On the other hand, the literature concerning the financing of small and medium enterprises (SMEs), as in many EU countries the largest part of the agricultural producers is indeed small farmers.

[7] A common, shared notion of green farming practices does not exist in literature or in the agricultural industry. In our analysis, we refer to this expression for any farming practice that may be entitled to be financed via green financing instruments. See further in the chapter on the issue of a lack of definition of green farming practices.

[8] As a matter of fact, to be useful, a proposed financing structure needs to have clear testable implications, so that the underlying paradigms may be supported or refuted by data. To this extent, the present work should be considered only a first step, setting up the theoretical foundations. Further empirical research will be needed to test the effectiveness of the discussed financing structures on the ground.

8.2.1 Farming Yields and Impact of Subsidies in the EU Agriculture

The weight of public support for European agriculture has been historically important. Even if successive reforms have progressively reduced the incidence of the CAP on the European budget,[9] the financing of agriculture in Europe still remains highly dependent on subsidies (see Table 8.1).[10] In this context, green farming practices have progressively gained attention. In the actual multiannual financial framework (2014–2020), the support for these practices is divided into two different budgetary envelopes. On the one side, funds are allocated as mandatory *greening* direct payments under the Pillar I.[11,12] On the other side, *agri-environment* and *climate* payments following the provider-gets principle are available under the Pillar II.[13]

As regards in particular the research on the financing of green farming practices at the European level, two main aspects convey a particular relevance in our study. The first concerns the determinants of the farmer's decision-making process. Consistently with the fact that different green farming practices have different effects on yields and land productivity (e.g. Wezel et al. 2014), it can easily be argued that the relation between

[9] The share of the CAP within the total EU budget has decreased sharply over the past 30 years despite the successive EU enlargements (from 73% in 1985 to 39% in 2013). Such a trend has been induced by a series of successive reforms, which have mainly had the objective of incentivising a progressive transition towards a more market-oriented system. Nonetheless, in the actual multiannual financial framework (2014–2020), the CAP funds amount to over € 55 billion per year (EC 2015b).

[10] Depending on the country, direct payments (hence excluding other forms of subsidy) may range from 15% or less (Cyprus, Lithuania, Malta, the Netherlands and Romania) to more than 40% (Ireland, Luxembourg, Slovakia and Sweden). Source of data: EC (2015b).

[11] In this paragraph, the citations in *italics* refer directly to the vocabulary used in the CAP provisions.

[12] To be eligible for mandatory *greening* payments, farmers have to comply with a number of practices considered beneficial for the environment. In particular, this refers to the maintaining of permanent grassland, crop diversification and the presence of an ecological focus area.

[13] *Agri-environment* and *climate* payments are considered within the Rural Development policy. These payments are cofinanced by the European budget and national or regional authorities, which have a large autonomy in designing their own multiannual programmes on the basis of the menu of measures available at the European level. The provider-gets principle states that farmers who sign up for environmental commitments beyond the reference level of mandatory requirements shall receive funds to cover the costs incurred and income forgone.

Table 8.1 EU farms output and subsidies (average values per farm)

	Total output (€)	*Total output/ total input*	*Total subsidies excluding on investments (€)*	*Subsidies on investments (€)*	*Environmental subsidies (€)*	*Balance current subsidies – taxes (€)*
Central Europe						
Austria	75,255	1.10	18,414	1342	6397	18,053
Belgium	265,975	1.15	24,745	2264	1804	23,328
France	195,887	1.01	29,893	1345	1300	27,759
Germany	266,707	1.06	34,821	794	3298	36,486
Luxembourg	192,653	0.93	44,445	15,875	9543	54,698
The Netherlands	490,248	1.12	16,875	666	1672	13,200
Mediterranean						
Croatia	23,200	1.06	4008	–	–	3471
Cyprus	40,769	1.18	4906	254	758	4801
Greece	21,783	1.27	6538	33	67	5847
Italy	52,951	1.41	6436	348	677	5092
Malta	39,675	1.25	2764	519	243	2712
Portugal	29,499	1.24	7057	1119	684	6784
Spain	52,181	1.30	9182	144	638	10,143
Northern Europe and UK						
Denmark	484,484	1.06	36,349	733	758	31,898
Finland	106,543	0.77	49,888	912	11,183	49,505
Ireland	69,754	1.05	20,276	232	2464	19,642
Sweden	199,885	0.89	40,395	–	10,538	40,215
United Kingdom	257,008	1.02	42,059	1365	7390	41,162
Eastern Europe and Baltic						
Bulgaria	38,872	0.96	10,675	482	1494	10,159
Czech Republic	344,709	0.89	95,128	3537	11,322	93,080
Estonia	111,296	0.90	25,063	4191	7386	24,808
Hungary	65,507	1.02	15,899	790	2444	14,932
Latvia	56,694	0.94	13,841	–	2105	13,400
Lithuania	42,555	1.08	10,024	1458	190	9641
Poland	31,390	1.15	5984	267	565	5589
Romania	12,967	1.49	2033	6	73	1858
Slovakia	609,681	0.78	164,039	6724	12,680	155,583
Slovenia	25,047	0.90	8088	1389	1726	7638
Total average	**70,346**	**1.11**	**11,101**	**420**	**1151**	**10,620**

(*continued*)

Table 8.1 (continued)

Notes: The table shows the main figures concerning the EU farms' output and the weight of the subsidies, including environmental subsidies. The latter are, on average, about 10.3% of the total subsidies (excluding subsidies on investments). Nevertheless, substantial differences exist country-wise in terms of impact of environmental subsidies over the total amount of subsidies (it can be observed a maximum of 34.7% for Austria and a minimum of 1.0% for Greece).

Source: authors' elaboration on data from Farm Accountancy Data Network (FADN). Data refers to 2013. The FADN sample covers approximately 80,000 holdings. They represent a population of about 5,000,000 farms in the EU, which covers approximately 90% of the total utilised agricultural area and accounts for about 90% of the total agricultural production

production output and the amount of subsidies represents a major decision factor. To this extent, it has been observed that little economic incentives may induce farms to opt out of even mandatory *greening* payments (Schulz et al. 2014).[14] However, financial compensation is a necessary but not sufficient condition to push farmers to undertake green farming practices. Literature has shown that non-financial factors may also be important (Siebert et al. 2006). In this respect, the possibility of maintaining a specific agricultural activity and preserving the management (Espinosa-Goded et al. 2010), the presence of certain qualitative attributes of the farmer and of the enterprise (Schulz et al. 2014) and the expected duration of the financing programme (Kuminoff and Wossink 2010) may contribute to the farmer's decision to be engaged in a public support scheme.

A second main issue typically linked to agricultural subsidies is regarding the inherent principal-agent problem and the incentive for farmers to cheat. In this regard, the operational characteristics of the programmes as well as the type of control and sanction systems in place are relevant. In particular, literature concerning the CAP has shown that *agri-environment* payments may provide incentives for payments based on cross-compliance,[15] while monitoring cross-compliance does not guarantee full respect of the other provisions (Bartolini et al. 2012). Concerning farmers' behaviour, it has been shown that farmers have incentives to cheat early over cheating late in the contract period on the basis of the differences in the expected

[14] These authors have, in particular, observed that specialised arable farms on highly productive land and intensive dairy farms are most likely to opt out of *greening* and renounce their entitlements.

[15] *Cross-compliance* is a mechanism that links payments to compliance by farmers with basic environmental and other standards. In the 2014–2020 multiannual financial framework, Pillar I and many Pillar II payments may be reduced in the case of non-compliance.

cost of compliance (Fraser 2012), and that farmers who have to face uncertainty in their production income are more likely to comply with the whole policy provision set as a means of risk management (Fraser 2002).

8.2.2 Features of the SME Financing

The issue of the financing of SMEs in Europe has become one of the key points of attention for policy makers and economists, in particular in the aftermath of the financial and economic crisis. Such consideration is justified as the SME segment represents an important portion of the value-added creation in Europe, especially in the non-financial sector.[16] To this extent, small size characterises agricultural producers in many European countries (see Table 8.2).

Literature has highlighted a number of features that characterise the access to external funds for SMEs. First, some authors have evidenced that SMEs are typically more financially constrained than large firms as a consequence of their limited access to alternative sources to bank lending (because of high transaction costs) and that such financial inability may represent a hurdle to their economic growth (e.g. Beck et al. 2009). Second, several works have shed light on the presence of credit limitations due to the opaqueness of the balance sheet and other relevant information that typically features SMEs and that create substantial asymmetries of information between borrowers and lenders. In this respect, limited information can affect lending from institutions that base their credit decisions principally on hard, objective and transparent data (e.g. Berger and Udell 2002), are characterised by complex hierarchical organisations (Stein 2002) and are physically distant from the potential borrowers (Bellucci et al. 2013). Third, empirical evidence also suggests that SMEs lending may be dropped by banks in favour of plain-vanilla types of lending, such as mortgages, in an attempt to limit the negative effects of opaque information (Liu et al. 2011). Finally, existing literature demonstrates that the quality of the relationship with the bank can play a role for SMEs in terms of cost of funding. The longer the relationship, the lower the loan rates and the fewer the loan covenants (Berger and Udell 1995).

[16]The SMEs segment represents 58% of the value-added creation and 67% of the employment in the non-financial sector in Europe. Data refers to 2014 (EC 2015a).

Table 8.2 EU farms financial position (average values per farm)

	Total assets (€)	*Total fixed assets (€)*	*Total fixed assets/total assets*	*Farm capital (€)*	*Farm capital/ total assets*
Central Europe					
Austria	452,770	357,816	0.79	354,264	0.78
Belgium	720,729	631,212	0.88	408,493	0.57
France	441,328	260,265	0.59	385,348	0.87
Germany	888,949	748,454	0.84	389,645	0.44
Luxembourg	1,151,439	976,202	0.85	663,998	0.58
The Netherlands	2,285,939	1,976,903	0.86	826,415	0.36
Mediterranean					
Croatia	154,886	141,972	0.92	92,182	0.60
Cyprus	179,583	144,176	0.80	81,863	0.46
Greece	108,009	103,258	0.96	57,973	0.54
Italy	389,804	281,063	0.72	172,262	0.44
Malta	194,903	180,331	0.93	111,899	0.57
Portugal	107,447	81,982	0.76	67,382	0.63
Spain	261,885	199,955	0.76	120,551	0.46
Northern Europe and UK					
Denmark	2,523,260	2,125,149	0.84	964,658	0.38
Finland	435,161	358,794	0.82	261,379	0.60
Ireland	926,583	866,954	0.94	191,327	0.21
Sweden	898,861	704,736	0.78	522,798	0.58
United Kingdom	1,807,977	1,635,705	0.90	427,585	0.24
Eastern Europe and Baltic					
Bulgaria	77,652	46,593	0.60	64,760	0.83
Czech Republic	985,969	742,030	0.75	838,356	0.85
Estonia	266,001	193,372	0.73	209,355	0.79
Hungary	172,167	105,732	0.61	129,403	0.75
Latvia	147,389	98,350	0.67	110,528	0.75
Lithuania	121,519	78,835	0.65	95,446	0.79
Poland	165,862	145,669	0.88	85,926	0.52
Romania	39,592	29,546	0.75	30,578	0.77
Slovakia	1,068,131	576,034	0.54	1,013,452	0.95
Slovenia	199,035	185,353	0.93	107,928	0.54

(*continued*)

Table 8.2 (continued)

Notes: Farm size can be estimated by the average total assets. The level of leverage, which can be to some extent a proxy of the possibility of accessing debt financing, can be approximated by the metric farm capital/ total assets (higher the value, lower the level of leverage). It can be observed that in counties such as Denmark, the Netherlands and United Kingdom a higher dimension of the firm is paired with a higher level of leverage. Conversely, in many Eastern-European countries, small dimension is paired with low debt. Concerning the first two economies in Europe, Germany and France, it can be observed that German farms are, on average, larger (about double in size) and more leveraged (almost doubly leveraged) than the French ones. For the latter, reduced size might lead to difficulties in accessing external financing

Source: authors' elaboration on data from Farm Accountancy Data Network (FADN). Data refers to 2013. The FADN sample covers approximately 80,000 holdings. They represent a population of about 5,000,000 farms in the EU, which covers approximately 90% of the total utilised agricultural area and accounts for about 90% of the total agricultural production

On the other hand, abundant evidence exists consistent with the idea that relationship lending[17] is more effective than transaction-based lending in limiting the information asymmetries between SMEs and financial intermediaries. In particular, relationship lending practices allow banks and other financial institutions to better collect and store soft information. Relationship lending institutions are hence prone to exploit this information over time by fostering a long-term connection with the borrower (e.g. Boot 2000; Berger et al. 2001). Similarly, another recognised feature of relationship lending institutions, and in particular of cooperative banks, is their resilience to monetary shocks in terms of lending supply. Literature linked to the *bank lending channel* shows that in periods of credit tightening cooperative banks are usually able to provide funds to their clients to a higher extent than other types of lending institutions (e.g. Bolton et al. 2013; Migliorelli 2018), even though some limitations to this countercyclical role may exist (Migliorelli and Brunelli 2017).

Despite these results, the conventional wisdom that large banks have little interest in serving SMEs has collected only ambiguous findings. Although many authors have found evidence that small and niche banks may better engage with SMEs through relationship lending using soft information, while large and foreign banks tend to lend less to SMEs (e.g. Mian 2006; Jimenez et al. 2009), such a view is not unanimous.[18] In

[17] Relationship banking can be defined as the provision of financial services by a financial intermediary to the market on the basis of both hard and soft information, the latter obtained through a long-term engagement and continuous interaction with the client (Cornée et al. 2018). Conventionally, cooperative banks and savings institutions are considered as practising relationship lending.

[18] Another factor that can induce small banks to focus on SMEs is the borrowers' concentration problem that they could suffer by lending to large enterprises.

particular, Berger and Udell (2006) argue that banks can strategically decide to use a set of different lending technologies (which include both transaction-based lending technologies[19] and relationship lending), and this choice may not be linked to the size of the bank. This can justify why, for example, the intensification of bank involvement with SMEs observed in various markets is neither led by small or niche banks, nor highly dependent on relationship lending (De la Torre et al. 2010).

8.3 Factors Hampering the Development of Green Finance in Agriculture

This section aims at identifying the main obstacles that still hamper the expansion of green finance in the EU agriculture today. To this extent, an approach similar to the one of Ashcraft and Schuermann (2008) and Migliorelli and Dessertine (2018) is adopted. These authors focused their investigation on the analysis of the nature of the relationships linking several actors involved in different financing structures.[20] Following such an approach, they were able to identify the key frictions able to harm the effectiveness (in terms of outreach) and the long-term sustainability (in terms of appropriate management of risks) of the financing structures under assessment. We mimic this approach in a more generic framework given by the financing of green farming practices in the EU agriculture. In this respect, we consider as a friction *any element potentially able to hinder the investment in green farming practices* via *green-labelled financing instruments.* Two main types of frictions can be identified: market-related frictions (stemming from possible market failures) and definition-related frictions. These frictions are reported in Fig. 8.3 and are discussed here below.

The Lack of a Definition of Green Farming Practices

Even though a number of provisions included in the CAP schemes identify activities (and related investments) that could be in principle considered

[19] Financial statement lending, asset-based lending, credit scoring and factoring are some of the most widely used transaction lending technologies.

[20] In more detail, Ashcraft and Schuermann (2008) analysed the specific agency problems at several points in the securitisation chain for subprime mortgages that ignited the Great Crisis, while Migliorelli and Dessertine (2018) focused on a possible securitisation mechanism to be introduced in the EU agriculture to foster environmentally friendly practices.

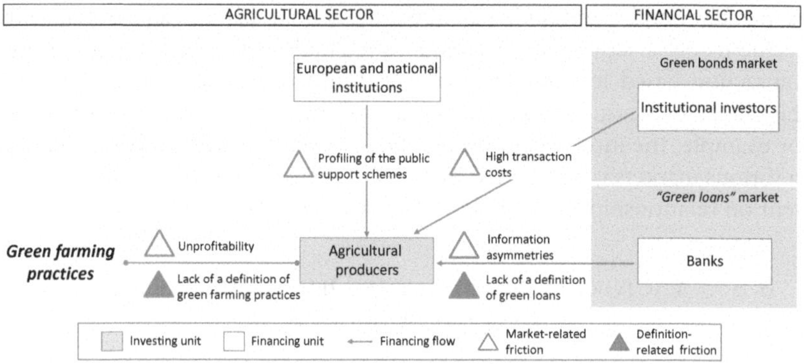

Fig. 8.3 Main frictions potentially able to hamper the investment in green farming practices via green-labelled financing instruments (Notes: Green loans market is in italic as no specific definition and structure of such a market currently exists). (Source: Author's elaboration)

green farming practices,[21] a specific definition or taxonomy for these practices does not currently exist. As a matter of fact, the lack of such a reference is a first important hurdle for the systematic development of green finance in EU agriculture. A specific, shared and widely applied definition of green farming practices or a detailed taxonomy would act as a facilitator for channelling investments, as it could be used as a reference by both farmers and financing institutions. Nonetheless, this definition issue is not specific to green farming practices in agriculture but it is indeed common to all sustainable activities. In this respect, EU institutions have already launched a specific initiative with the aim to progressively introduce an EU classification system.[22]

[21] These include the provisions linked to the mandatory *greening* direct payments and the *agri-environment* and *climate* payments following the provider-gets principle already mentioned in paragraph 8.2.1.

[22] The initiative of the European Commission launched in 2018 and aimed at establishing an EU classification system (or taxonomy) for sustainable activities is a step in this direction. This initiative should result in a Report of the Commission technical group providing a taxonomy for climate mitigation activities by Q1 2019 and a taxonomy on climate change adaptation and other environmental activities by Q2 2019. This taxonomy will not be included in legislative measures. For a more detailed discussion, see in particular Chaps. 6 and 7.

High Transaction Costs to Access the Bonds Market

Abundant evidence exists demonstrating that the incidence of the transaction costs on the capacity of a firm to access financial markets increases with a decrease in the size of the firm (e.g. Beck et al. 2009). According to this evidence, large agricultural producers may face only limited or no constraints in issuing green securities, in particular *green bonds*, to finance their green farming projects.[23] Nevertheless, the largest part of the agricultural producers in many European countries consists of small farmers, with no direct access to the bond market and with no other concrete forms of financing other than bank lending (which may also be subject to limitations).[24] Hence, a strong argument should be made according to the idea that one of the main causes of the limited role of green-labelled finance in agriculture has to be found in the small size that typically features agricultural producers and their inability to issue green securities on the debt market. It seems indeed reasonable to explain a (even relevant) part of the observed negligible share of *green bonds* issuance in agriculture in line with this argumentation.

Information Asymmetries Between Agricultural Producers and Banks

Literature has proved that the lack of reliable information from potential borrowers can induce financial intermediaries, in particular banks, not to grant credit. This lack of information is usually exacerbated in the case of opaque SMEs (e.g. Berger and Udell 2002), which is often the case in the agricultural market. In this regard, a lower debt exposure for small farms vis-à-vis larger ones seems confirmed by data in Europe (see Table 8.2). As a matter of fact, many small and medium agricultural producers over Europe may suffer from a certain level of underfunding and this can impact (also, but not only) the investment in green farming practices. The opaqueness of the information system that may feature many small agricultural producers over Europe should be hence considered another possible impediment to the expansion of green farming financing, in particular as concerns the crucial aspect of the total volumes financed.

[23] *Green bonds* may suffer less than other types of financing instruments of the lack of a definition or taxonomy of green activities. This is mainly due to the development of specific methodologies and labels by the financial industry.

[24] This may be particularly so in many East-European countries and, to a lesser extent, in France (see again Table 8.2).

The Lack of a Definition of Green Loans

Partially linked to the previous point, it should be still considered that a fair assessment of the effective role of green finance in agriculture as directed via the banking channel is today hindered by the lack of a specific, shared and widely applied definition of *green loans*.[25] As a matter of fact, it is likely that a certain share of banking loans in the portfolios of banks would be in line with criteria on the use of proceeds that could eventually characterise a *green loan*. However, today such criteria are not fully developed or applied, and loans effectively financing green farming practices are not reported as *green loans*. Even though the practice of funding of green activities through traditional loans is indeed common to many industries, it is heavily widespread in the agricultural sector. This is again a consequence of the presence of a large portion of SMEs with no effective financing alternative than banking financing.[26] In point of fact, a formal recognition of *green loans* (eventually backed by a structure definition or a taxonomy for green farming practices, as already discussed) could streamline and eventually encourage the banks' lending activity in this field. Possibly, a definition of *green loans* could also be the basis for further policy initiatives aiming at encouraging the financing of green activities and investments.[27]

[25] A first initiative to define green loans is dated March 2018 when the Loan Market Association (LMA) published the first Green Loans Principle (GLPs), mainly mimicking the Green Bond Principles (GBPs) initiative. In the framework established by the LMA, green loans are defined as "*any type of loan instrument made available exclusively to finance or re-finance, in whole or in part, new and/or existing eligible green projects*". To be recognised as such, the concerned organisations also have to align with the four core management components of the use of proceeds, process for project evaluation and selection, management of proceeds and reporting (LMA 2018).

[26] Such (apparently secondary) a problem may have indeed important policy implication. In fact, it may be expected that a policy intervention defining green farming practices in agriculture (e.g. through a specific taxonomy), and conversely constituting a basis for identifying green financing instruments, may conduct to a wide re-labelling of existing bank loans in green loans, with no significant effects in terms of additionality. For policy initiatives aiming at increasing the financing flow towards green farming practices to be effective, complementary actions would indeed be needed (e.g. in the form of fiscal or regulatory incentives for banks to issue green loans to otherwise not financed projects). See Chap. 7 for a more in-depth discussion on this issue.

[27] In this respect, a promising movement in research, supported by policy makers, is emerging trying to link sustainability risks and financial risks (e.g. by considering that sustainable investments may reduce the frequency and the incidence of catastrophic natural events, thus also reducing their economic consequences). In the long term, and as a measure to further encourage sustainable finance, regulatory provisions for a different prudential treatment of the exposures related to sustainable projects might be introduced to the benefit of financial

Unprofitability of Green Farming Practices and Profiling of the Public Support Schemes

Existing literature clearly shows that different farming practices may have different levels of farming integration and yields, and hence the profitability condition may not hold for all the possible investments. As a matter of fact, farmers may be pushed to drop the green farming option as economically not viable (e.g. Schulz et al. 2014; Wezel et al. 2014). Nevertheless, this behaviour may produce an undesirable outcome for social welfare.[28] To minimise this problem, an argument should be made in favour of allowing direct payments or other types of subsidies to farmers in order to support their income when carrying out green farming practices. In this way, the entity of the market failure (in terms of farmers' income loss or unprofitable market prices) can be compensated by public spending. In Europe, this sustenance principle is generally established and managed through a number of schemes stemming from the CAP provisions. However, it can be easily argued that the CAP gives today little or no emphasis to green finance. This is mainly due to the actual profiling of public support schemes, which do not specifically foresee green-labelled financing instruments.[29] In this respect, no specific provisions are in particular dedicated to the establishment of ad hoc financing structures aiming at fostering green farming practices by allowing institutional investors (e.g. banks, insurance companies or pension funds) to buy green-labelled securities. As a matter of fact, the possibility for this kind of investors to contribute in financing green assets would represent an important step forward in the development of green finance in agriculture.[30,31]

intermediaries (e.g. in case of more favourable requirements in terms of capital absorption for loans issued to finance sustainable projects). In the EU perspective, a *supporting factor* is already foreseen in the European Commission plan for sustainable finance (EC 2018). See also Chap. 6.

[28] To this extent, also note that positive externalities linked to environmentally friendly practices are typically not considered in the farmer's individual investment choice pattern.

[29] This attitude can be explained by the traditional "incremental" evolution of the CAP over time (which inherits at every revision a relevant part of the support schemes used in the past) and by the relative novelty of the green finance discipline (which then requires time to be integrated in a complex policy scheme).

[30] In addition, the effective involvement of institutional investors could result in a more efficient use of public spending. If used in blending instruments (e.g. in the form of a guarantee), public support could become an effective means of leverage for the overall investment levels (in terms of ratio between public participation and the inflow of private resources).

[31] In this respect, relevant financial theory suggests that, in order to be more oriented towards the development of green finance, the public support should be kept in the form of

8.4 The Potential Role of Dedicated Financing Structures

This section presents the key features of three possible financing structures whose introduction may contribute to accelerate the development of green finance in agriculture by softening the incidence of some of the frictions previously analysed. In particular, the analysis is focused on the possibilities given by securitisation, guarantee funds and funds of funds. The financing structures assessed are market-oriented but backed by some form of public support. This support is indeed appropriate as directed to minimise the effects of the market failures that may exist in the intersection between the agricultural market and the financing markets and that can result in financing limitations for agricultural producers. We refer in particular to the high transaction costs to directly access financial markets and the possible information asymmetries featuring the relationship between agricultural producers and banks. To this extent, the public support embedded in the financing structures should be considered additional to the one given through grants in order to compensate farmers to undertake unprofitable green farming practices. On the other hand, the financing structures analysed do not aim and are not able to effectively contribute to reduce the incidence of the definition-related frictions on the development of green finance in agriculture, which was previously discussed.

However, the added value of the introduction of a specific financing structure to face a market failure should not be taken for granted. This value is indeed directly proportional to the entity of the market failure and to the capacity of the financing structure to overcome it (in terms of effective outreach of final beneficiaries—in this case farmers—and by means of its technical implementation). For example, in markets in which the average size of the agricultural producers is relevant and no particular impediments to the access

grants or direct payments only when aimed at compensating agricultural producers from carrying out unprofitable but green farming practices. Otherwise, it should be embedded in market-oriented financing structures (e.g. by means of guarantee) allowing a wide range of investors to take part in the financing of green farming practices. A first step in this direction has been recently made with the introduction of ad-hoc financing instruments in the framework of the European Funds for Strategic Investments (EFSI). In particular, since 2015, it is possible to create financing instruments combining funds from the EFSI and funds from the European Agricultural Fund for Rural Development (EAFRD), within the Rural Development (RD) policy. One of the declared objectives of these initiatives is to create a leverage effect with the public funds used (in terms of the ratio between the total amount of the resources mobilised by the financing instrument and the amount of the public support).

to the debt market exist, financing structures aiming at streamlining the link between farmers and financial markets are unlikely to add significantly to the financing perspectives of the potential beneficiaries (as no market failures may even exist).[32] Conversely, the same financing structures can be expected to exploit their full policy utility in markets in which the farmland is disperse and farmers face substantial hurdles to access finance. For these reasons, an assessment of the market conditions and of the technical fit of a financing structure in facing a specific market failure should represent an essential prerequisite for any policy initiative of this type.[33]

8.4.1 Securitisation

A standard securitisation process allows illiquid assets (e.g. mortgages, loans, short-term credit) owned by one or more financial institutions to be pooled and transferred to an ad hoc vehicle which issues tradable securities backed by those assets. These securities can be transacted in the secondary market. The originating entity is hence cashed out, while the securities issued by the vehicle are repaid through the cash flow of the original assets. Usually, an external credit enhancer is used to mitigate the risk for the final investors and create highly rated financial instruments.[34] The increased provision of liquidity is one of the obvious advantages of the securitisation at a single-bank level and one of the most important determinants of the decision to securitise assets (e.g. Cardone-Riportella et al. 2010; Farruggio and Uhde 2015). In particular, existing studies argue that a true sale securitisation[35] leads to an effective recalibration of the bank's balance sheet composition by disposing of illiquid assets and injecting cash (e.g. Gorton and Pennacchi 1995). The management of capitalisation through the

[32] In such a case, financing structure backed by public support could even go against the state aid regulation in EU.

[33] This practice is indeed already adopted by EU policy makers in several programmes, through the so-called *ex-ante assessments* (see for example the *Methodological handbook for implementing an ex-ante assessment of agriculture financial instruments under the EAFRD*).

[34] In this regard, excess spread (the practice of issuing notes with an overall yield lower than that of the underlying assets) and overcollateralisation (the practice of issuing an amount of notes lower than the available underlying assets) are also used as sources of internal credit enhancement and to cover transaction costs linked to the securitisation operation.

[35] In a true sale securitisation, the ownership of the underlying exposures is transferred or effectively assigned to a securitisation special purpose entity. In contrast, in a synthetic securitisation, the underlying exposures are not transferred, but the related credit risk is transferred by means of a guarantee or derivative contracts.

transfer of risk and the regulatory arbitrage (e.g. Affinito and Tagliaferri 2010) and the realisation of profits opportunities (e.g. Cardone-Riportella et al. 2010; Ahn and Breton 2014) have also been identified as bank-level determinants for launching securitisation operations.

More importantly for our analysis, literature has observed that potential systemic advantages of securitisation lay in the possibility for a wide range of investors to access asset classes traditionally reserved for retail banks. In more detail, this may be the case of SME lending and short-term commercial papers (Duffie 2008; IMF 2015). Likewise, consolidated literature has highlighted the contribution of securitisation to the diversification of the risk along the whole intermediation chain (Allen and Carletti 2006). The pooling and tranching processes allow the cash flow from the underlying assets to be restructured and profiled. This allows, on the one side, to mitigate idiosyncratic risk inherent to single-level loans and, on the other side, to create securities classes able to appeal to a large base of investors with different risk appetite (Caballero and Krishnamurthy 2009). Finally, literature has proved that securitisation may be an effective means to stimulate loan supply. In particular, this happens when banks use the capital freed after the sale of the assets to accept new credit risk (Altunbas et al. 2009).[36]

Securitisation can be an instrument able to foster green finance in agriculture, in particular, in contexts in which the average farmers' size is small and significant difficulties exist for farmers to access the financial markets. Nevertheless, to be effective, a standard securitisation mechanism requires

[36] Despite these advantages, existing literature highlights certain risks linked to securitisation practices. To this extent, a number of works have analysed the effects of the information asymmetries and the moral hazard that may feature the relationship between originator and final investors. In fact, banks and other financial institutions may tend to accept reducing their credit standards and transfer the risk to the market. In this respect, evidence has been documented, in particular for the subprime mortgages in the United States, which have been accused of triggering the financial crisis in 2007. Based on these studies, the absence of *skin in the game* has been the basis of a misalignment in the incentives between originators and final investors. This phenomenon has eventually caused a sensitive reduction in the quality of the underlying assets (Keys et al. 2009; Mian and Sufi 2009). For this reason, all recent regulation proposals on securitisation have included risk retention clauses concerning the originator. In more detail, the provision of maintaining a minimum nominal value of the first-loss tranche or of each of the tranches sold or transferred to investors is constantly proposed to limit opportunistic behaviours (BCBS 2014; IMF 2015; EC 2015c).

specific and interconnected characteristics. First, it should be in the form of an *originate-and-distribute* model,[37] allowing for only new loans to be considered for securitisation in order to increase the possibility to produce effective *additionality*. Second, the public support should consist of an integrated programme foreseeing both eligibility checks for loans to be securitised (in particular to assure that only green farming practices are financed) and a general guarantee embedded in the securitisation structure. The latter should be of the amount necessary to provide notes appealing to institutional investors in terms of risk-return. Finally, the notes issued by the securitisation vehicle would benefit to be labelled as *green securities* in order to steer the demand in the financial market.[38] Figure 8.4 shows a simple securitisation mechanism of this type.

The expected benefits of the introduction of this financing structure should be found in the possibility for institutional investors to (indirectly) finance green farming practices and, as concerns agricultural producers, in the expected increased availability of funds and/or in lower funding costs.[39] On the other hand, the biggest limitations of the use of securitisation are linked to the complexity of its technical implementation and to the often high operational costs.[40]

[37] In an *originate-and-distribute* model, originators (typically banks) issue new loans with the intention to successively securitise them. Nevertheless, in order to avoid opportunistic effects, a characterising feature of the financing structure should be that originators maintain a certain level of *skin in the game*. In other words, they must keep in their balance sheets a quota of the first-loss tranche or a quota of all the tranches issued by the vehicle.

[38] For a more detailed analysis on how an *origination-and-distribute* securitisation mechanism can be introduced in agriculture, see Migliorelli and Dessertine (2018).

[39] The expected reduction in the funding cost for agricultural producers is directly linked to the incidence of the public guarantee on the risk-profile of the notes issued by the securitisation vehicle.

[40] In addition, securitisation has been blamed for contributing to the explosion of the subprime mortgage crisis in the United States and igniting the financial contagion worldwide. In particular, the negative view was due to the observation that securitisation had probably incentivised lax credit policies and poor asset quality standards. For these reasons, in the aftermath of the financial crisis, securitisation operations have registered historically low levels of issuance both in Europe and in the United States. Policy makers and international organisations have recently reacted by proposing amendments to existing regulations in an attempt to contrast the misalignments observed in the securitisation chain and give new impulsion to the market. In Europe, the ongoing reform aims, in particular, to identify criteria for simple, transparent and standardised securitisation (STS).

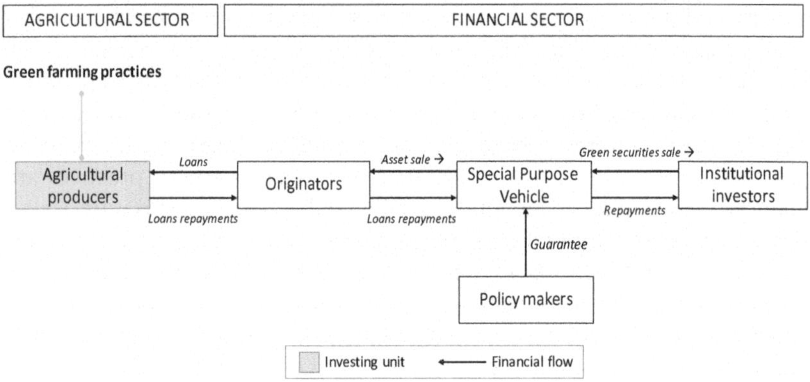

Fig. 8.4 A simple securitisation scheme in the agricultural sector. (Source: Author's elaboration)

8.4.2 Guarantee Funds

Guarantee funds are another family of financing structures that could be used to encourage the development of green finance in agriculture. Guarantee funds are usually set by policy makers when small firms cannot gain adequate access to credit, or to credit on equally favourable terms when compared to large firms of equal risk (Cowling and Siepel 2013). These types of financing structures hence seek to provide loan supply security to those small firms that would otherwise be unable to obtain debt finance through conventional means, in particular, through the banking channel (Riding et al. 2007). To do that, the guarantee embedded in these financing structures has the main function to encourage banks to issue credit. In operational terms, and in order to increase the probability of effective ***additionality***, the policy programmes backed by a guarantee fund usually limit their scope to newly issued banking loans directed to finance projects with specific characteristics that otherwise would not be financed. To this extent, ***eligibility*** criteria are usually defined and used to screen perspective investment projects. For standard loan guarantee funds, the repayments of the loans release the corresponding proportion of the guarantee and may eventually free up this amount for reinvestment.[41]

[41] Critics of these financing structures usually highlighted the problems associated with low levels of ***additionality*** (even new loans awarded under these schemes might have been obtained without support), high administration costs and an often limited leverage effect (calculated by the ratio of total loans issued on the amount of public support).

In order to be considered financing structures able to contribute to foster green finance in agriculture, guarantee funds should be hence built foreseeing specific ***eligibility*** criteria that would restrict the investments financed to green farming practices only.[42] As concerns the key aspect of the risk covered by the guarantee, three limitations may be adopted to bind the use of resources to the effective needs and to avoid the moral hazard and possible opportunistic behaviours from the originators (which could otherwise tend to issue loans to earn the origination fee and then transfer the entire credit risk to the guarantor). First, the guarantee can be restricted in amount (the so-called *capped amount*). Second, this amount should represent a predefined (maximum) portion of the total portfolio of loans foreseen by the fund (the *guarantee cap rate*). Third, the guarantee should cover only a portion of the value of each loan (the *guarantee rate*).[43,44] Figure 8.5 summarises the functioning of a guarantee fund to be deployed in the agricultural sector.

Nevertheless, when it comes to the analysis of the effective impact of these financing structures on the development of green finance in agriculture, some limitations can be observed. These limitations are linked to the fact that guarantee funds do not allow institutional investors to take position in green assets, with the latter remaining in the balance sheet of the originators (i.e. banks). As a consequence, the leverage effect of the public intervention (calculated as the ratio between the inflow of private resources and the amount of the public guarantee) remains somewhat limited and substantially lower than other financing structures open to the secondary market.[45]

[42] As already underlined, a definition of green farming practices (e.g. through a specific taxonomy) will ease this task.

[43] Another differentiating feature of a guarantee fund, seldom used, could be the provision to entirely cover the so-called *first loss* that is the entire amount of the first defaulted loans. In these cases, particular attention should be given to opportunistic behaviours of the originators.

[44] The setting of capped amount, guarantee cap rate and guarantee rate practically determines the size of the portfolio of loans covered by the guarantee. This can be calculated as: *Size of the portfolio (EUR) = capped amount (EUR)/guarantee cap rate (%)/guarantee rate (%)*. The size of the portfolio of loans object of the guarantee has to be established on the basis of a risk assessment taking into account the specific market conditions.

[45] The leverage ratio for initiatives in the EU of guarantee funds can be expected to be between 4.0 and 6.0. As a comparison, the expected leverage ratio for financing structures backed by securitisation mechanisms may reach 15.0.

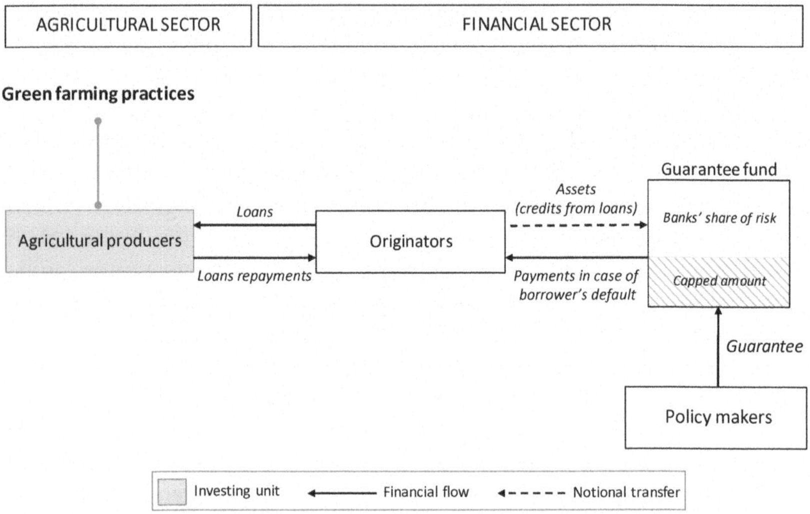

Fig. 8.5 Simple guarantee fund scheme in the agricultural sector (Notes: The flow of assets from the originators to the guarantee fund is notional as there is no real transfer of ownership). (Source: Author's elaboration)

8.4.3 *Funds of Funds*

A third category of financing structures that can be used to foster green finance in agriculture is represented by the so-called funds of funds. Even if these structures are quite flexible in terms of financing instruments directed to final beneficiaries (both debt and equity) and in terms of types of beneficiaries to be reached (both SMEs and large companies), when it comes to policy-driven interventions they are used in particular to channel equity investments to small and medium businesses. In more detail, these financing structures generally consist of a fund that, on the one side, invests in the equity of venture capital and private equity funds (or similar investment agents) operating in specific sectors, and, on the other side, may issue on the market tradable securities to finance its activity.[46] The final investment is usually directed towards innovative and fast-growing firms and is normally limited in time (i.e. the funds sell their quotas in the

[46] This second feature, even though it can increase the overall size of the fund by opening to market contribution, is not mandatory.

firms once it has reached the expected growth potential). When the fund of funds is established by a national or international financial organisation to face existing market failures, it may be backed by a public guarantee.

Funds of funds can bring significant advantages in terms of green farming development, but they face some relevant limitations. The main advantage of a fund of funds is that it can foster the equity investment in innovative, high-potential small and medium agricultural companies that otherwise would risk to be capped in terms of growth potential. In this respect, the possibility to receive direct capital infusion may represent a significant enhancement in the investment possibilities of the firm.[47] In addition, being in the form of equity, this support normally does not alter the level of leverage and improve the risk-profile of the firm.[48] In practical terms, the fund of funds needs to identify suitable intermediary investment bodies (e.g. venture capital or private equity firms) active in the agricultural sector and agree with the latter on an investment strategy. The most important limitation of these financing structures is that they are fully suitable only for financing innovative, high-potential firms in a relatively early stage of growth[49] and are not appealing for more traditional producers with more mature businesses. This characteristic restricts the potentiality of this instrument to a relatively small (but critical) segment of the green farming sector. On the liability side, when prompted to foster green farming investments, the fund of funds might issue *green securities*[50] to collect funds on the market and increase its overall investment potential. To this extent, the presence of a public guarantee may contribute to reduce the risk-profile of the fund and attract institutional investors. Figure 8.6 shows a simple fund of funds scheme to be deployed in the agricultural sector.

[47] In some cases, venture capital and private equity funds can also enter in the management of the firm, potentially bringing benefits in terms of additional competencies and market knowledge.

[48] In some industries, the entrance of venture capital and private equity firms is the capital of a company often accompanied by the issuance of new debt.

[49] This is linked to the investment strategy of the venture capital or private equity funds that are remunerated by the sell-off of their quota when the firm has reached the expected growth potential.

[50] Even in this case, a standard definition of green securities issued by funds of funds is currently not developed or in use.

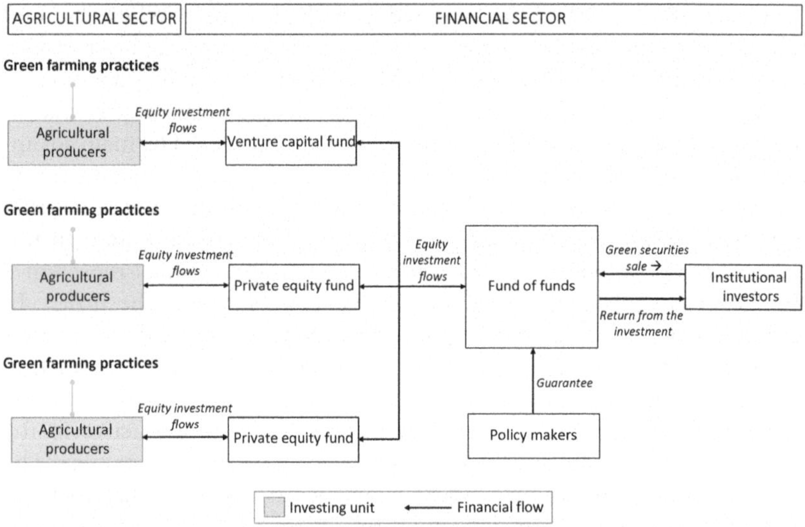

Fig. 8.6 Simple fund of funds scheme in the agricultural sector (Notes: Equity investment flows include both the initial equity investment in the firm from the venture capital or private equity fund and the return from exiting this investment (by selling the participation in the firm, usually in the market). The return from the investment due from the fund of funds to institutional investors depends on the structuring of the fund and the type of financing instruments issued by the fund of funds (equity and/or debt)). (Source: Author's elaboration)

8.5 The Possible Role of the Cooperative Sector

In this section we analyse the potential role of the cooperative sector in fostering green farming practices. In this respect, it should be first mentioned that the cooperative paradigm[51] can be applied in a variety of structures and governance types in the agriculture sector as well as in the

[51] In this respect, it is important to underline the main peculiarities of the cooperative paradigm. Primarily, it is reflected in a unique ownership structure. Cooperatives cannot exclude new members unless motivating the reasons and, most importantly, the *one-head-one-vote* rule is in use in the decision-making processes. Furthermore, cooperatives have a very limited profit-seeking nature. In fact, most of them face constraints in terms of profit distribution. Finally, the link with the territory and the mutualism principle mainly steers the cooperative activity. Normally, it has to be focused first of all towards their members and in the territory where they mainly operate.

financial sector (e.g. Bijman and Iliopoulos 2014). With regard to the farmers' cooperatives, sensitive differences in size and participation have been observed (e.g. Gijselinckx and Bussels 2014). In some cases, agricultural cooperatives or federations of agricultural cooperatives have been able to reach a nation-wide scale and have been structured to embrace activities in other sectors. In Europe, the average cooperatives' market share in selling agricultural products is around 40%, and above 50% in countries such as Austria, France, Ireland, the Netherlands and the Scandinavian countries.[52] As in agriculture, the cooperative model is widely used in the financial intermediation sector. Financial cooperatives have been historically created with the scope to respond to the financing needs of their members. To this extent, cooperative banks in particular have reached a systemic importance in some key European countries such as Austria, France and Germany (e.g. Cornée et al. 2018).[53] Given this framework, the notion of the cooperative intermediation we use is somewhat blurred. In the analysis, we generally refer to a cooperative operational entity and a cooperative financial entity. In more detail, we identify the cooperative operational entity as a union of farmers in a specific territory. Its main scope is to market the agricultural output. Nevertheless, other activities may be included, such as coordination of the production factors or farmers' representation. On the other hand, with cooperative financial entity we denote a structure whose scope is to provide credit and other financing services to its members. In doing so, they mainly use relationship lending.

8.5.1 Relationship Lending and Other Lending Technologies

As we have already observed, because of the expected information asymmetries characterising the relationship between farmers and banks, a concrete risk of market failure and significant crowding-out effects may occur with respect to small farmers. To deepen the analysis on that point (and even before considering the introduction of specific financing structures), it can be stated that the entity of market failure also depends on the type of lending

[52] See Bijman and Iliopoulos (2014).

[53] In some cases, cooperative banks have experienced exceptional growth. As a consequence, those financial institutions have reached high levels of hybridisation and have been substantially transformed into universal banks (e.g. Crédit Agricole in France or Rabobank in the Netherlands).

technology used by banks (Berger and Udell 2006). In this respect, any lending technology consists of a series of methods and instruments applied in each of the phases that typically compose the credit process: contact generation, borrower's needs analysis, information collection, borrower's creditworthiness analysis, contract formalisation, credit issuance and credit monitoring.[54] It can be easily argued that the financing of green farming practices conveys a significant distinctiveness in the credit process. In this respect, an argument can be made concerning the fact that relationship lending would be an effective means to reduce information asymmetries and streamline at least the first three phases of the process (e.g. Boot 2000; Berger et al. 2001) and that this effect would even be amplified in the case of lending institutions specialising in the agriculture sector. In this respect, cooperative banks are by far the most important relationship banking institutions.[55] Relationship lending is the main lending technology they use contrary to the traditional transaction lending technologies largely adopted by commercial banks in particular (Cornée et al. 2018).

8.5.2 Cooperative Intermediation

The key contribution of the cooperative sector in a specific financing structure can be hence expected in the loans origination phase. In particular, the cooperative operational entity could facilitate the credit process by providing first-instance financial counselling services to farmers. In this context, the administrative officer of the cooperative operational entity acts as a contact point between farmers and the loans officer of the cooperative financial entity. On the one hand, the administrative officer collects information on the financing schemes available, presents them to farmers and helps filling in loan documentation. On the other hand, s/he introduces potential clients to the loans officer, illustrates the farmers' financing needs and has a limited negotiation power. In the cooperative intermediation scenario, both the administrative officer and the loans officer are hence the repositories of the soft information concerning

[54] For more details, see Migliorelli and Dessertine (2018).

[55] In many instances, banking institutions that follow the same organisational pattern (e.g. a decentralised decision-making) may exhibit quite a similar behaviour in their lending practices. This is especially the case for other stakeholder-oriented banks such as community banks and savings banks.

potential borrowers.[56] It can be argued that the contribution of this intermediation chain to the reduction of the expected market failure affecting SMEs is twofold. First, part of the file costs would be centralised and absorbed by the cooperative operational structure. To this extent, the financial entity would bear almost no costs in the phases of contact generation, borrowers' needs analysis and information collection. Second, the soft information stored by the loans officer thanks to relationship lending could be used to improve the effectiveness of the borrower's creditworthiness analysis. As a consequence, in the cooperative intermediation scenario, an increase in the marketability of the SME lending could be expected. For financing structures backed by loans (such as securitisation and guarantee funds), the presence of the cooperative sector can hence represent a significant facilitating factor.[57,58]

[56] The notion of a repository of the soft information we refer to is the one discussed by Berger and Udell (2002).

[57] Nevertheless, it should be argued that the systemic relevance of this issue has to be evaluated country-wise. In fact, both the composition of the farming industry and of the banking sector (the latter in particular in terms of the presence of transaction lending technologies specialising in SMEs lending) may play a significant role in determining the final marketability of the loans and their average amount. In countries in which the average farm size is higher and the access to debt for farms is easier, the transaction-based banking channel could be expected to be fairly effective in financing green farming practices. This can be the case for countries such as Denmark, the Netherlands or the United Kingdom. On the other hand, in countries in which the farmland is dispersed and agricultural firms rely on the personal wealth of the farmer or the family more than on debt, there is the likelihood of a reduced outreach while using transaction-based lending increases. This may be particularly so in many East-European countries and, to a lesser extent, in France (see Table 8.2).

[58] In the cooperative intermediation scenario, some inherent risks may exist in the case of an explicit or implicit hierarchical link between cooperative operational entity and cooperative financial entity. Such a situation may occur, in particular, in the case of large agricultural cooperatives or federations of cooperatives controlling a financial institution with the aim of serving their members. In such a case, conflicts of attribution may materialise between the operational and the financial entities. Furthermore, a dominant operational entity would tend to impose laxer credit standards and reduce the monitoring of the borrowers. In the mid-term, this would probably produce a deterioration of the quality of the loans issued and securitised. To limit systemic risks due to the transfer of the loans to the market through securitisation, a rigid governance structure assuring the decisional independency of the financial entity would have to be put in place and clearly communicated to the market. As a matter of fact, this is the typical case of captive financial institutions operating within larger industrial or commercial groups and that might be replicated in the largest agricultural cooperatives or federations of agricultural cooperatives.

8.6 Concluding Remarks

Green finance still plays a limited role in the EU agriculture. In this chapter we observed how its development has been so far hindered by sector-specific factors such as the presence of a large number of small farmers (implying limited access to the debt market—e.g. to issue *green bonds*—and possible credit limitations when financing via the banking channel), the actual profiling of the public support (which does not specifically foresee or encourage the use of green securities) and the lack of a definition for green farming practices and of *green loans* (which may refrain intermediaries from recognising and eventually streamlining their green financing activity). We have also argued that the introduction of specific financing structures can contribute to at least partially overcome this situation. To this extent, securitisation, guarantee funds and funds of funds can represent actionable solutions to foster green finance in agriculture. Nevertheless, their effectiveness must be assessed on the basis of the specific market conditions in the country or in the region in which they are intended to be introduced, and no *one-fits-all* solutions exist. Finally, we have observed how the cooperative sector can play a role in improving the efficiency and the effectiveness of the credit process on the basis of some of the above-mentioned financing structures by smoothening the borrower-lender relationship.

Stemming from these analyses, an argument can be made according to the idea that a structured policy intervention would be necessary today to allow green finance to be developed in agriculture at least at the same extent as it has been developed in other industries. In fact, market forces will hardly be able to overcome the obstacles arising from the inherent characteristics of the sector. This policy intervention should foresee at least two coordinated actions. First, it should aim at identifying a non-controversial definition of green farming practices and, conversely, a basis for categorising green finance instruments in agriculture. This action may (should) be included in a larger one to consolidate a definition of green activities at a systemic level, for example, through a complete taxonomy (an initiative in this direction has already been tabled by the European Commission).[59] This action will also ease the adoption of a definition of *green loans* in the financial industry. Second, the policy intervention would benefit from foreseeing an adjustment of the profiling of the public

[59] A wider discussion on these issues is given in Chap. 6.

support given through the CAP. In this respect, two reference principles should be followed. On the one hand, public support should be kept in the form of grants or other forms of direct payments when compensating agricultural producers to carry out unprofitable but green farming practices. On the other hand, public support should be embedded in specific financing structures aiming at both easing the access to financial markets for all types of agricultural producers and allowing private market investors (in particular institutional investors) to take stable position in financing green farming practices. In doing so, and in order to steer the market demand, specific attention should be made to the establishment of green labels (or labelling criteria) for the securities issued by these financing structures.[60]

References

Affinito, M., & Tagliaferri, E. (2010). Why do (or did?) banks securitize their loans? Evidence from Italy. *Journal of Financial Stability, 6*, 189–202.

Ahn, J. H., & Breton, R. (2014). Securitization, competition and monitoring. *Journal of Banking and Finance, 40*, 195–210.

Allen, F., & Carletti, E. (2006). Credit risk transfer and contagion. *Journal of Monetary Economics, 53*, 89–111.

Altunbas, Y., Gambacorta, L., & Marques-Ibanez, D. (2009). Securitisation and the bank lending channel. *European Economic Review, 53*, 996–1009.

Ashcraft, A. B., & Schuermann, T. (2008). *Understanding the securitization of subprime mortgage credit*. Staff Reports 318, Federal Reserve Bank of New York.

Bartolini, F., Gallerani, V., Raggi, M., & Viaggi, D. (2012). Modelling the linkages between cross-compliance and agri-environmental schemes under asymmetric information. *Journal of Agricultural Economics, 63*(2), 310–330.

Basel Committee on Banking Supervision (BCBS). (2014, December). *Revisions to the Basel securitisation framework*.

Beck, T., Demirgüç-Kunt, A., & Martínez Pería, M.S. (2009). *Banking SME around the world: Lending practices, business models, drivers and obstacles*. World Bank Policy Research Working Paper, 4785.

[60] In terms of standards and labels for green financial products, the European Commission has mainly focused its attention on green bonds, with a Report of the Commission technical expert group on a standard for green bonds expected by Q2 2019. In addition, an assessment of applying the EU Ecolabel to financial products is also ongoing as of Q2 2018. Nevertheless, green labelling criteria for securities issued by structured finance vehicles (e.g. securitisation or funds of funds) are not yet under specific discussion. A wider analysis of these issues is given in Chap. 6.

Bellucci, A., Borisov, A., & Zazzaro, A. (2013). Do banks price discriminate spatially? Evidence from small business lending in local credit markets. *Journal of Banking and Finance, 37*, 4183–4197.

Berger, A. N., & Udell, G. F. (1995). Relationship lending and lines of credit in small firm finance. *The Journal of Business, 68*(3), 351–381.

Berger, A. N., & Udell, G. F. (2002). Small business credit availability and relationship lending: the importance of bank organizational structure. *The Economic Journal, 112*, F32–F53.

Berger, A. N., & Udell, G. F. (2006). A more complete conceptual framework for SME finance. *Journal of Banking and Finance, 30*, 2945–2966.

Berger, A., Klapper, L., & Udell, G. (2001). The ability of banks to lend to informationally opaque small businesses. *Journal of Banking and Finance, 25*, 2127–2167.

Bijman, J., & Iliopoulos, C. (2014). Farmers' cooperatives in the EU: Policies, strategies, and organization. *Annals of Public and Cooperative Economics, 85*(4), 497–508.

Bolton, P., Freixas, X., Gambarotta, L., & Mistrulli, P. E. (2013). *Relationship and transaction lending in a crisis*. BIS Working Papers, No 417.

Boot, A. W. A. (2000). Relationship banking: what do we know? *Journal of Financial Intermediation, 9*, 7–25.

Caballero, R. J., & Krishnamurthy, A. (2009). Global imbalances and financial fragility. *American Economic Review, 99*(2), 584–588.

Cardone-Riportella, C., Samaniego-Medina, R., & Trujillo-Ponce, A. (2010). What drives bank securitisation? The Spanish experience. *Journal of Banking and Finance, 34*, 2639–2651.

Cornée, S., Fattobene, L., & Migliorelli, M. (2018). An overview of cooperative banking in Europe. In M. Migliorelli (Ed.), *New cooperative banking in Europe. Strategies for adapting the business model post-crisis*. London: Palgrave Macmillan.

Cowling, M., & Siepel, J. (2013). Public intervention in UK small firm credit markets: Value-for-money or waste of scarce resources? *Technovation, 33*(8), 265–275.

de la Torre, A., Soledad Martínez Pería, M., & Schmukler, S. L. (2010). Bank involvement with SMEs: Beyond relationship lending. *Journal of Banking & Finance, 34*(9), 2280–2293.

Duffie, D. (2008). *Innovations in credit risk transfer: Implications for financial stability*. BIS Working Paper, 255.

Espinosa-Goded, M., Barreiro-Hurlé, J., & Ruto, E. (2010). What do farmers want from agrienvironmental scheme design? A choice experiment approach. *Journal of Agricultural Economics, 61*(2), 259–273.

European Commission (EC). (2015a, November). *Annual report on European SMEs 2014/2015*.

European Commission (EC). (2015b, March). *CAP post-2013: Key graphs and figures.*

European Commission (EC). (2015c, September). *Proposal for a regulation of the European Parliament and of the Council laying down common rules on securitisation and creating a European framework for simple, transparent and standardised securitization.*

European Commission (EC). (2018, March). *Actionplan: Financing sustainable growth.* COM (2018) 97 final.

Farruggio, C., & Uhde, A. (2015). Determinants of loan securitization in European banking. *Journal of Banking and Finance, 56*, 12–27.

Fraser, R. (2002). Moral hazard and risk management in agri-environmental policy. *Journal of Agricultural Economics, 53*(3), 475–487.

Fraser, R. (2012). Moral hazard, targeting and contract duration in agri-environmental policy. *Journal of Agricultural Economics, 63*(1), 56–64.

Gijselinckx, C., & Bussels, M. (2014). Farmers' cooperatives in Europe: Social and historical determinants of cooperative membership in agriculture. *Annals of Public and Cooperative Economics, 85*(4), 509–530.

Gorton, G. B., & Pennacchi, G. G. (1995). Banks and loans sales marketing non-marketable assets. *Journal of Monetary Economics, 35*, 389–411.

International Monetary Fund (IMF). (2015, January). *Securitization: The road ahead. IMF Staff discussion note.*

International Panel on Climate Change (IPCC). (2014). *Climate change 2014. Mitigation of climate change.*

Jimenez, G., Salas, V., & Saurina, J. (2009). Organizational distance and use of collateral for business loans. *Journal of Banking and Finance, 33*, 234–243.

Keys, B. J., Mukherjee, T., Seru, A., & Vig, V. (2009). Financial regulation and securitization: Evidence from subprime mortgage loans. *Journal of Monetary Economics, 56*, 700–720.

Kuminoff, N. V., & Wossink, A. (2010). Why isn't more US farmland organic? *Journal of Agricultural Economics, 61*(2), 240–258.

Liu, M., Margaritis, D., & Tourani-Rad, A. (2011). Asymmetric information and price competition in small business lending. *Journal of Banking and Finance, 35*, 2189–2196.

Loan Market Association (LMA). (2018). *Green loans principles.* London.

Mian, A. (2006). Distance constraints: The limits of foreign lending in poor economies. *Journal of Finance, 61*, 1465–1505.

Mian, A., & Sufi, A. (2009). The consequences of mortgage credit expansion: evidence from the 2007 mortgage default crisis. *Quarterly Journal of Economics, 124*(4), 1449–1496.

Migliorelli, M. (Ed.). (2018). *New cooperative banking in Europe. Strategies for adapting the business model post-crisis.* London: Palgrave Macmillan.

Migliorelli, M., & Brunelli, S. (2017). The transmission of the monetary policy in the euro area: The role of the banks' business model. *Annals of Public and Cooperative Economics, 88*(3), 302–322.

Migliorelli, M., & Dessertine, P. (2018). Time for new financing instruments? A market-oriented framework to finance environmentally friendly practices in EU agriculture. *Journal of Sustainable Finance and Investment, 8*(1), 1–25.

OECD. (2017). *Mobilising bond markets for a low-carbon transition*. Paris.

Riding, A., Madill, J., & Haines, G. (2007). Incrementality of SME loan guarantees. *Small Business Economics, 29*(1–2), 47–61.

Schulz, N., Breustedt, G., & Latacz-Lohmann, U. (2014). Assessing farmers' willingness to accept "greening": Insights from a discrete choice experiment in Germany. *Journal of Agricultural Economics, 65*(1), 26–48.

Siebert, R., Toogood, M., & Knierim, A. (2006). Factors affecting European farmers participation in biodiversity policies. *Sociologia Ruralis, 46*, 318–340.

Stein, J. (2002). Information production and capital allocation: Decentralized versus hierarchical firms. *Journal of Finance, 57*, 1891–1922.

Wezel, A., Casagrande, M., Celette, F., Vian, J., Ferrer, A., & Peigné, J. (2014). Agroecological practices for sustainable agriculture. A review. *Agronomy for Sustainable Development, 34*, 1–20.

CHAPTER 9

Fintech, Digitalization and Blockchain: Possible Applications for Green Finance

Gregor Dorfleitner and Diana Braun

9.1 Introduction

The Financing Gap

To reach the Sustainable Development Goals (SDG), Blakstad and Allen (2018) report an annual finance gap of USD 2.5 trillion with USD 1 trillion related to renewable energy. Similarly, Fuessler et al. (2018) report that an average of USD 3.5 trillion per year needs to be invested in the energy sector until 2050 to achieve the two degrees goal of the Paris Agreement. Hence, there is wide consensus on the existence of a huge gap in green investments, whereas public finance cannot provide sufficient capital and private investment and capital have to be increasingly mobilized (Fuessler et al. 2018; Plunkett et al. 2016). Under the term *green* or *sustainable* finance we include financial instruments, investments, measures or flows of finance towards green projects, companies or technologies, which make a contribution to the environment and/or the mitigation of climate change.

G. Dorfleitner (✉) • D. Braun
University of Regensburg, Regensburg, Germany
e-mail: gregor.dorfleitner@ur.de

M. Migliorelli, P. Dessertine (eds.), *The Rise of Green Finance in Europe*, Palgrave Studies in Impact Finance,
https://doi.org/10.1007/978-3-030-22510-0_9

Barriers to the Mobilization of Green Finance
The mobilization of green finance is facing several barriers which can be clustered in four key categories:

- First of all, green projects face difficulties in finding sources of funding and oftentimes lack access to traditional sources of finance (e.g. banks, institutional investors) (Lam and Law 2016; Plunkett et al. 2016). Green projects are usually located in the energy sector and are based on innovative technologies, requiring elaborate R&D and high upfront cost (Lam and Law 2016). In addition, investors' risk-return expectations (Plunkett et al. 2016) or strict requirements (e.g. collateral to apply for a bank loan) (Lam and Law 2016) are not met due to uncertainty in commercialization (Lam and Law 2016), future cash flows or profitability, etc. (Bonzanini et al. 2016).
- Second, mobilizing funding and flow of financing is further slowed down due to intermediaries involved in a centralized system. Fuessler et al. (2018) argue that in traditional financial systems finance flows towards green initiatives involve burdensome "bureaucratic processes" and multiple stakeholders, which leads to higher transaction cost and hampers private investment.
- Third, there are no universal and broadly accepted standards or definitions to back the reliability of existing green labels (Blakstad and Allen 2018; Neves and Prata 2018). This can relate to green or sustainable labels for financial instruments, like green bonds, where standards such as the green bond principles serve as leading frameworks for green certification but where there is currently no common definition for "greenness" (Shishlov et al. 2016).
- Fourth, and closely related to the third category is the lack of transparency and accountability (Neves and Prata 2018). This includes transparency on and access to (impact) data and the capability to measure, report and verify whether an investment target is *green* as well as the tracking and evaluation of the use of proceeds and their additionality (Fuessler et al. 2018; Neves and Prata 2018).

The lack of universal standards and definitions (Blakstad and Allen 2018; Neves and Prata 2018) coupled with lack of transparency and accountability (Neves and Prata 2018) leads to a prevalent threat of greenwashing, which negatively impacts the flow of finance towards green projects.

Regarding these challenges, the following sections of this chapter aim at analysing how fintech and blockchain application can contribute to the mobilization of green finance. In Europe, several initiatives are already shaping discussions on fintech and blockchain to mobilize green finance. *Crowdfundres*,[1] for example, is an initiative that advances crowdfunding for renewable energy projects and provides policy recommendations regarding regulatory frameworks. The *Climatechain*[2] is a French research project that analyses the potential of blockchain technology in effectuating the Paris Climate Agreement. In 2018, the *European Commission* has published two action plans: the "Fintech action plan for a more competitive and innovative European financial sector" (European Commission 2018b) and the "Action plan for financing sustainable growth" (European Commission 2018a). However, according to CRIC (2018), the connection between fintech and sustainability is lacking.

Overview on Fintech and Blockchain Applications in Green Finance
We argue that fintechs offer a valuable opportunity to disrupt the financial system and to mobilize green finance. Castilla-Rubio et al. (2016) state that the realization of climate pledges requires a "reset of the global financial system", in which fintechs act as "core disruptor". Financial innovations such as crowdfunding, robo-advisors or blockchain increasingly offer green investment opportunities for the general public (Blakstad and Allen 2018) and make the financial system easier accessible and more efficient (Castilla-Rubio et al. 2016). Based on the definition of Dorfleitner et al. (2017) we cluster fintechs into financing, asset management and blockchain to discuss possible applications in green finance (see Fig. 9.1).

In a nutshell, robo-advisors, crowdfunding and blockchain technology show the potential to address the barriers of mobilizing green finance (see Fig. 9.2). They facilitate access to new sources of finance, especially from private investors. The mobilization is additionally supported by offering decentralized systems, bypassing traditional intermediaries such as banks or other financial institutions, decreasing costs and inefficiencies. Blockchain technology further enables effective monitoring, reporting and verification. However, each of these applications still requires adequate legal and regulatory frameworks as well as uniform standards and definitions.

[1] See www.crowdfundres.eu as source (access date 2019-03-04) and for additional information.

[2] See www.theclimatechain.org as source (access date 2019-03-04) and for additional information.

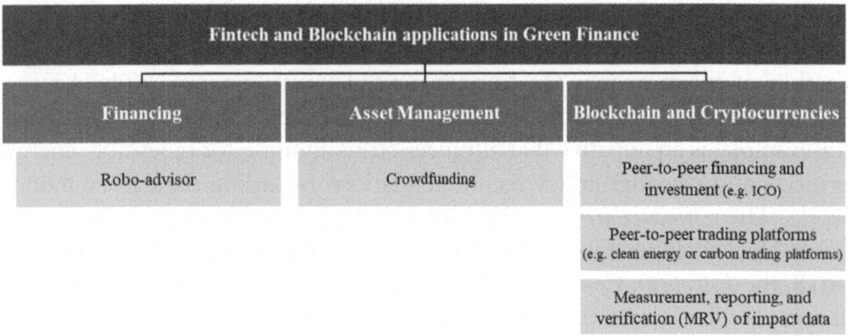

Fig. 9.1 Fintech and blockchain applications in green finance. (Source: Authors' elaboration)

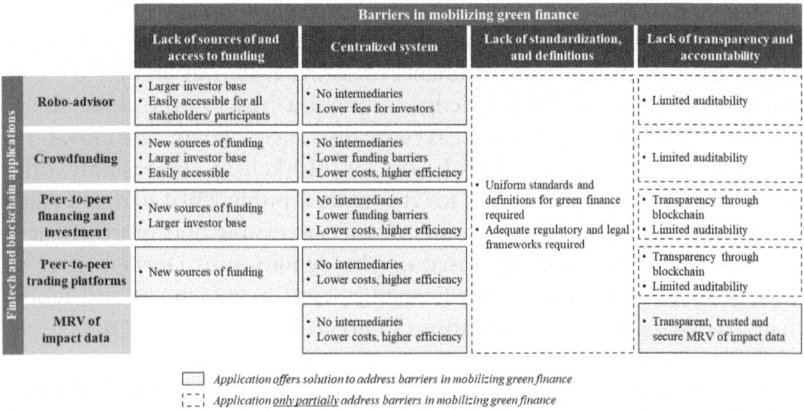

Fig. 9.2 How applications of fintech and blockchain address barriers in mobilizing green finance. (Source: Authors' elaboration)

9.2 Fintech Applications in Green Finance

9.2.1 Green Robo-Advisors

Definition and Functioning of Robo-Advisors

Robo-advisors are fintechs disrupting traditional asset management with low-cost investment advice and portfolio management services (Citi 2016; Kaya 2017). They enable individuals not having the financial means for costly investment advice to obtain a new form of such advice (Castilla-Rubio et al. 2016). These fintechs provide automated investment advice and portfolio management services using computer algorithms to find optimal investment strategies (Dorfleitner et al. 2017; Kaya 2017). Suggested investment strategies are based on the answers investors provided during the "initial investor screening" (Kaya 2017), in which they are asked questions related to their investment objective, investment amount, their risk appetite, product preferences and so on. Robo-advisors can invest in various funds but predominantly invest in ETFs (Dorfleitner et al. 2017). Compared to traditional asset managers, robo-advisors charge lower fees (Citi 2016) and have lower minimum investment requirements, making them especially suitable for younger people with lower wealth (Fisch et al. 2018).

We define green robo-advisors as those platforms that either are solely dedicated to sustainable investing or offer—among standard portfolios—the option of selecting sustainable investing. Following Global Sustainable Investment Alliance (GSIA) (2016) sustainable investing "is an investment approach that considers environmental, social and governance (ESG) factors in portfolio selection and management". In the following we will use the terms *green*, *sustainable*, *socially responsible* or *responsible* interchangeably as description for portfolios that follow an investment approach based on ESG criteria. In Europe, robo-advisors, such as *VisualVest* or *LIQID* offer sustainable investing besides standard portfolios, whereas the German robo-advisor *Vividam* solely focuses on sustainable investments (see Table 9.1).

How can robo-advisors ensure that a portfolio modelled for an investor can be labelled as green? So far, there is no global standard or uniform process, how investments have to be evaluated or screened to be categorized as sustainable. The green robo-advisors *Vividam*,[3] *VisualVest*,[3] *LIQID*,[3] *Nutmeg*[3] and *Wealthify*[3] are mainly assessing green portfolios along ESG criteria. They usually apply several strategies such as excluding funds not complying with ESG criteria (negative screening) and evaluating the remaining along ESG criteria following the MSCI SRI index (e.g. *LIQID*[3]). Further standards and criteria in the analysis are, for example, Sustainable Development Goals (SDG) of the United Nations or Principles

[3] See Table 9.1 for an overview of robo-advisors. For more detailed information see the respective webpages stated in Table 9.1.

of Responsible Investing (PRI) as well as robo-advisors' individual criteria. Hence, despite some standards applied in the sustainability analysis, the overall process, the application and weighting of criteria to label portfolios as green is individually determined by each robo-advisor.

With green portfolios, customers usually have to pay higher fund or product costs compared to standard portfolios. For example, the robo-advisors *VisualVest*, *LIQID*, *Nutmeg* and *Wealthify* charge higher fund costs for green portfolios (see Table 9.1). In the case of *LIQID*, Fromm (2018) states that the higher fees result from the costly process of measuring and evaluating sustainability criteria which require, for example, high-quality data, and due to the rather low number of funds or ETFs accomplishing sustainability criteria.

Robo-advisors have experienced a significant boom in recent years. In Europe, the robo-advisor segment has increased from USD 7365 million assets under management (AuM) in 2017 to USD 30,052 million in 2019 and is expected to amount USD 97,626 million AuM by 2022 (Statista.com 2019). So far, there is no data on the volume of sustainable assets managed by robo-advisors. However, social and environmental impact investing is of increasing importance for the so-called Generation Y, which also shows a stronger preference for digital financial innovations like robo-advisors (Fisch et al. 2018; Ramos et al. 2016). Combined with the fact that by 2030 they will make up half of total AuM (Ramos et al. 2016), one can presume that robo-advisors with a dedicated sustainable investment option are of increasing importance.

Benefits

Regarding the above-mentioned barriers hampering the mobilization of green finance, robo-advisors come along with several beneficial features that offer a powerful solution to overcome these barriers (see Fig. 9.2).

Robo-advisors offer access to investment opportunities for a broader investor base, including new, less wealthy and less sophisticated investors (Kaya 2017; O'Keefe et al. 2016) who have not been served by traditional asset management (Dorfleitner et al. 2017). Green robo-advisors hereby represent a source of mobilizing private capital and investment. Higher accessibility (O'Keefe et al. 2016) and increased "financial inclusion" (Kaya 2017) is enabled through lower fees and charges (Fisch et al. 2018; O'Keefe et al. 2016) as well as a lower minimum investment requirement compared to traditional asset managers (Fisch et al. 2018). Furthermore,

Table 9.1 Green robo-advisors in Europe

	Vividam	*VisualVest*	*LIQID*	*Nutmeg*	*Wealthify*
Website	www.vividam.de	www.visualvest.de	www.liqid.de	www.nutmeg.com	www.wealthify.com
Location	Germany	Germany	Germany	UK	UK
Offering	Solely offer ethical and ecological investment strategies	"GreenFolios"—portfolios from sustainable funds (besides standard investment offering[a])	Service category "Global Impact" for sustainable investing	Socially responsible investing (besides standard investment offering[b])	Ethical investing (besides standard investment offering)
Product	Investment in active funds (no use of ETFs)	ETFs and actively managed investment funds	Passive investing: ETFs	Passive investing: ETFs	Passive investing: mutual funds and ETFs
Sustainability criteria	ESG criteria SDG criteria	Sustainability analysis of funds based on ESG criteria and results of MSCI ESG research; Scoring of GreenFolios (e.g. along ESG Quality Score)	Negative screening, and scoring of remaining companies against ESG criteria (based on MSCI SRI)	Negative screening (following MSCI SRI and MSCI ESG); Scoring of each portfolio (SRI and non-SRI) against ESG factors	Positive and negative screening based on ESG scores; Cooperation with ethical fund providers (PRI)
Min. investment	EUR 75 per month with EUR 3500 min. lump sum payment	EUR 500	EUR 100,000	GBP 500 GPB with GBP 100 (monthly)	No min. investment
Annual management fee	1.08%	0.6%	0.5%[c] (depending on amount of investments)	0.35–0.75%[d] (depending on amount of investments)	0.4–0.7% (depending on amount of investments)
Other fees	*Average product cost*: 1.36–1.76% (depending on investment strategy)	*Average fund cost*: 0.63–1.55% (GreenFolios) 0.31–0.45% (VestFolios – standard portfolios)	*Average product costs*[c]: 0.41% (sustainable portfolios) 0.15% (standard portfolios)	*Average fund cost*[b]: 0.33% (SRI portfolios) 0.19% (non-SRI portfolios)	*Average fund cost*: 0.54% (ethical portfolios) 0.22% (non-ethical portfolios)

Notes: This table only gives an overview of robo-advisors in Europe with a sustainable investment offering. The list is not comprehensive and does not raise any claim to completeness

Sources: Authors' elaboration on webpages of respective platforms (access date: 2019-02-11)

[a]VisualVest has two categories of investment: GreenFolios (sustainable portfolios) and VestFolios (non-sustainable/standard portfolios)

[b]Nutmeg offers four ways of investing: general investment, stocks and shares ISA, stocks and shares Lifetime ISA and personal pension. Focus here is only on general investment. Within each category one can choose between three investment styles, where one is the socially responsible investment style. The other two styles (fully managed and fixed allocation) do not consider SRI (www.nutmeg.com)

[c]We compare fees for LIQID Global (standard/non-sustainable portfolios) and LIQID Global Impact (sustainable portfolios).

[d]We only compare the fees for socially responsible and fully managed portfolios, since both are pro-actively managed.

ease of application (e.g. opening a portfolio on a robo-advice platform within 15 minutes (Kaya 2017)) and improved user experience (O'Keefe et al. 2016) with user-friendly web/mobile applications (Kaya 2017) also positively impact accessibility. Positive user experience is further supported by the possibility of "goal-based" investing (Ramos et al. 2016) where such goals can include social or environmental values. Robo-advisors also address the above-mentioned challenges coming along in a centralized financial system, since they do not require intermediaries as with traditional asset management involving, for example, financial advisors charging high fees. Hence, with green robo-advisors green and sustainable investing will be easily accessible and realizable for a broad investor base, especially for private people.

Limitations and Challenges

In order to model green portfolios, robo-advisors need to screen investments to assess their *greenness*. Standards such as the ESG criteria, offer a trustworthy means of orientation and confidence for investors. However, the above-mentioned prevalent lack of uniform standards and definitions for green investments, coupled with a lack of transparency, accountability and monitoring capabilities still is present with robo-advisors. In an article published in the Financial Times, Beioley (2019) argues that there are varying definitions across investment providers on what are ethical or sustainable portfolios. In addition, Beioley (2019) remarks that according to the platforms IG, Moneyfarm and Scalable Capital the volume of sustainable ETFs is still too small to achieve a sufficient spread for a well-diversified portfolio being a reason for them not to offer such portfolios.

Furthermore, robo-advisors cannot offer the same level of individualized advice as human financial advisors (Fisch et al. 2018), because standardized questionnaires for initial investor screening only provide limited information on investors' specificities such as risk aversion (Fein 2015). Modelled portfolios might also lack accuracy regarding investors' preferences (Fisch et al. 2018). Another challenge is the increasing competition in the robo-advisor segment, which will require robo-advisors to move towards a greater product diversification (Fisch et al. 2018). Differentiation with a green investment offering could hereby represent an opportunity.

9.2.2 Green Crowdfunding

Definition of Crowdfunding and Different Crowdfunding Models
Crowdfunding is a financing mechanism that enables entrepreneurs, small businesses or projects to raise funding from a large number of contributors connected via the crowdfunding platform (Belleflamme et al. 2014). Each contributor—which can be any individual—typically only needs to invest a small amount, in contrast to traditional financing which is raised from a small group of sophisticated investors (Belleflamme et al. 2014). The crowdfunding platform hereby replaces traditional intermediaries like banks, which often represent a barrier to access financing, especially for small and innovative businesses (Belleflamme et al. 2014). In addition, crowdfunding platforms have lower fixed and transaction costs compared to financial institutions, representing cost savings for entrepreneurs and investors (Lam and Law 2016).

There are four different types of crowdfunding, whose main difference is related to the way investors are refunded for their investment (Dorfleitner et al. 2017). In the donation-based crowdfunding projects are funded via donations from the investors (Bradford 2012). In the reward-based crowdfunding investors are rewarded in form of non-monetary consideration (Dorfleitner et al. 2017). In the crowdlending model investors provide funding in form of loans and in return receive the repayment of their principal and possibly additional interest payments (Bradford 2012). In the equity crowdfunding model investors are rewarded for their contributions being offered a stake in the profits or returns of the funded project (Bradford 2012).

This categorization also holds for green crowdfunding, however, according to Lam and Law (2016) depending on the scale and time frame of the project different crowdfunding models are more suitable. First of all, they state that donation- and reward-based crowdfunding are applicable for initial phases of a project whereas crowdlending and equity crowdfunding are more suitable means of financing at later stages to further expand the project. Second, they argue that for "small-scale renewable energy" or "green innovation projects" donation- and reward-based crowdfunding are adequate for fundraising, where "large-scale renewable energy projects" should rather set on crowdlending or equity crowdfunding, since their capital requirements and planning horizon are significantly larger. Furthermore, according to Butticè et al. (2019) green projects in general have higher funding goals compared to other crowdfunding campaigns.

There are multiple crowdfunding platforms that specialize in funding of green energy projects (see Table 9.2). However, despite the increasing importance of crowdfunding as a source of capital for the mitigation of climate change, the scale is still small (Adhami et al. 2017; Dilger et al. 2017). Large crowdfunding platforms lack explicit categories for green or sustainable projects (Leblanc 2017). In addition, Butticè et al. (2019) state that on Kickstarter, one of the largest global crowdfunding platforms, only 9.5% of all the projects[4] can be categorized as green initiatives.

Benefits
Several barriers in mobilizing green finance can be addressed with crowdfunding (see Fig. 9.2). Green crowdfunding facilitates green projects in finding sources of finance and getting access to a broader investor base including private investors. It serves as a suitable means of finance for green projects, such as renewable energy projects which are characterized by "high up-front costs" (Lam and Law 2016), since investors can step in with a rather low investment amount, which makes them less inhibited to provide funding for such projects. Crowdfunding platforms also open up the opportunity to access a global investor base.

In addition, funding barriers are lower for green projects (Bonzanini et al. 2016) since crowdfunding enables the decentralized collection of funding without relying on traditional intermediaries. This reduces costs (e.g. transaction costs) usually involved in a central system with intermediaries (Bonzanini et al. 2016). In addition, risks of investors are reduced, since they are shared among several small investors (Bonzanini et al. 2016; Vasileiadou et al. 2016), with each investor only accounting for a relatively small amount of funding. Furthermore, no trust in large institutional intermediaries is required, which reduces counterparty risks (Galen et al. 2018; Neves and Prata 2018). Finally, crowdfunding enables the financing of (smaller) projects that otherwise would not get access to financing (Bonzanini et al. 2016) due to strict criteria such as high investment capital requirements (Vasileiadou et al. 2016) or fixed costs of due diligence activities (Bonzanini et al. 2016).

By its capabilities of mobilizing and bringing together a significant amount of people, crowdfunding is not only a means of financing a project

[4] Butticè et al. (2019) are analysing projects within the period of July 1, 2009 and July 1, 2012.

Table 9.2 Green crowdfunding platforms in Europe

Platform	*Website*	*Location*	*Platform type*	*Min. investment*	*Total amount invested*	*Since*
Oneplanetcrowd	www.oneplanetcrowd.com	Netherlands	Sustainable crowdfunding	Defined by entrepreneur	>30 Mio. EUR	2012
Abundance	www. abundanceinvestment.com	UK	Sustainable crowdfunding (green energy, energy efficiency, housing)	5 EUR	~88 Mio. GBP	2012
Lumo	www.lumo-france.com	France	Renewable energy crowdfunding	25 EUR	>5 Mio. EUR	2012
Enerfip	www.enerfip.fr	France	Renewable energy crowdfunding (solar, wind, hydro, biomass)	10 EUR	>15 Mio. EUR	2014
Bettervest	www.bettervest.com	Germany	Crowdfunding platform for sustainable energy projects	50 EUR	>10 Mio. EUR	2012
Econeers	www.econeers.de	Germany	Renewable energy crowdfunding	250 EUR	>10 Mio. EUR	2013
Trine	www.jointrine.com	Sweden	Crowdfunding platform for solar energy	25 EUR	>14 Mio. EUR	2015
Ecrowd!	www.ecrowdinvest.com	Spain	Crowdfunding platform for sustainability: categories energy, water, health, education, etc.	50 EUR	~5 Mio. EUR	2013

Sources: Authors' elaboration on webpages of respective platforms www.crowdfunding.de and www.crowdcircus.de (access date: 2019-02-21); (Bonzanini et al. 2016; Caneva and Alonso 2018; Leblanc 2017)

but also a means of intensifying the social awareness, involvement and support for green projects (Bonzanini et al. 2016; Vasileiadou et al. 2016). Furthermore, societal support and commitment can then translate into political support, creating a positive impact on increasing investments and initiatives towards climate action (Vasileiadou et al. 2016).

Green crowdfunding platforms provide transparency for investors. Investors get an overview on green investment options with easy access to information on projects (e.g. objectives, use of proceeds, impact) and fast subscription processes (Vasileiadou et al. 2016). In addition, investors are receiving regular updates on status, progress and success of the funded projects (see e.g. green crowdfunding platform *Oneplanetcrowd*[5]).

Challenges and Limitations

Green crowdfunding clearly has the potential to access new sources of investment pushing forward finance flows towards climate action, mobilizing at the same time political and societal engagement. However, green crowdfunding still faces the challenge of insufficient transparency and accountability as well as lacking standards and definitions. For investors it is still difficult to comprehend and ascertain the accountability of the funded project as well as to measure the impact the project creates regarding sustainability. Alonso et al. (2017) argue that the due diligence of crowdfunding projects and respective companies is low and that there exists a lack of trustworthiness and credibility. Lam and Law (2016) also state the possibility of fraud with limited due diligence and verifiability of information compared to traditional investments. In addition, there are no uniform standards or definitions that determine which crowdfunding projects can be labelled as green or standardized criteria projects have to fulfil to be categorized as *green*. Each crowdfunding platform can individually decide which projects they classify as *green*. Information asymmetry between investors and entrepreneurs (Bergmann et al. 2016) is another prevalent risk, where investors are usually non-professional, without comprehensive understanding of business-related criteria or risks (Vasileiadou et al. 2016) and are influenced in their decisions by "herding behaviour" of the crowd (Adhami et al. 2017). Moreover, the technology of green projects (e.g. renewable energy) is innovative and still in its infancy with a higher likelihood of failure (Lam and Law 2016). Hence, crowdfunding

[5] See Table 9.2 for an overview of crowdfunding platforms. For more detailed information see the respective webpages stated in Table 9.2.

projects come along with a non-negligible risk exposure (Vasileiadou et al. 2016) and low investor protection at the same time (Adhami et al. 2017). Other risks related to crowdfunding are cyber insecurity (Lam and Law 2016) and lack of a common legal framework in Europe (Bergmann et al. 2016).

To increase the legitimacy and importance of crowdfunding as an adequate means of financing sufficient investor protection has to be provided which requires governmental support and regulation (Bergmann et al. 2016; Bonzanini et al. 2016; Vasileiadou et al. 2016) in a way that it fosters the scaling up of crowdfunding in Europe but does not hamper innovation (Caneva and Alonso 2018).

Table 9.2 gives an overview on existing green crowdfunding platforms in Europe. However, this list is not comprehensive and does not raise any claim to completeness. Its objective is to give an indication on the quantity and variety of existing green crowdfunding platforms in Europe. Another form of crowdfunding based on blockchain technology will be introduced in the next section. In order to set the scene, we will first give a short introduction on blockchain before elaborating on respective applications in green finance.

9.3 Blockchain Applications in Green Finance

9.3.1 Blockchain in a Nutshell

Definition of Blockchain

Blockchain is a distributed ledger that permanently, immutably and transparently records and stores transactions across a peer-to-peer network (Citi 2016; PwC 2018). Transactions have to be verified by each node of the network, are compiled in one block and added to the existing chain of blocks in a permanent and immutable way authenticated by cryptographic signatures (Castilla-Rubio et al. 2016; PwC 2018). The network can verify transactions without relying on a central authority and provides a “single source of truth” (Citi 2016). Regarding the verification process, there are two mechanisms used to create consensus in the network: the proof-of-work algorithm (e.g. used with Bitcoin), which requires significant computational power and high energy consumption, and the proof-of-stake algorithm (e.g. used with Ethereum) requiring a less complex verification process (PwC 2018).

Benefits of Blockchain Technology

Based on the features stated by Galen et al. (2018), Neves and Prata (2018) and PwC (2018), we summarize five key technical features of blockchain that build the basis to unlock the mobilization of green finance and to overcome respective barriers (see Fig. 9.2).

(1) *Transparency:* Blockchain allows anyone with access to the network to see the state of operated transactions and data at any point in time where each transaction is validated trough the network (Galen et al. 2018; Neves and Prata 2018). This creates transparency and improves monitoring, reporting and verification of transactions and data (Galen et al. 2018; PwC 2018).

(2) *Immutability:* Each transaction added to a block is time-stamped, verified and immutably stored where any changes to already entered data would immediately be detected by the network (Neves and Prata 2018). The database is not stored centrally or maintained by a central authority but distributed across the network (distributed ledger technology) (Citi 2016; PwC 2018). This is beneficial regarding trusted and secure measurement, collection and evaluation of impact data, such as data along the supply chain of products to evaluate their carbon footprint.

(3) *Authentication and Identity management:* Blockchain gives anyone access to a unique identity to digitally sign transactions via a unique public and private key (Galen et al. 2018). This makes (financial) transactions more reliable, more efficient and easier to process at lower cost (Neves and Prata 2018). It also enables the financial inclusion of minorities (e.g. unbanked) (Galen et al. 2018; Neves and Prata 2018), allowing anyone to access green investment independent of the geography. These features are beneficial to access new sources of capital from a broader investor base and to facilitate green finance flows.

(4) *Trusted and secured peer-to-peer transactions without intermediary:* Blockchain is a decentralized system enabling trusted and secured peer-to-peer transfer of value between unknown parties where trust is generated through the network itself based on consensus algorithms and smart contracts but without relying on a central authority (Neves and Prata 2018; PwC 2018). This reduces transaction costs, administrative costs and counterparty risks (Galen et al. 2018) as well as processing time (PwC 2018). Hence, blockchain facilitates peer-to-peer financial transactions towards climate action.

(5) *Tokenization and cryptocurrencies:* Closely related to the feature of peer-to-peer transactions are cryptocurrencies and tokens serving as transferable value. Blockchain enables to tokenize, for example, carbon emissions or other natural assets, which can then be easily traded, for example, for other cryptocurrencies as well as real currencies in exchange for cryptocurrencies (Blakstad and Allen 2018). This facilitates new forms of financing and investment of green projects such as blockchain-enabled crowd financing via initial coin offerings (ICO). In addition, it opens up the opportunity to develop peer-to-peer trading platforms, for example, for clean energy.

Challenges and Limitations

There are four key challenges regarding the applicability and implementation of blockchain technology: high energy consumption, technological restrictions (e.g. scalability, interoperability), usability (suitable for mainstream application) and legal and regulatory frameworks (Fuessler et al. 2018; Neves and Prata 2018; PwC 2018).

One of the key limitations of blockchain technology is the enormous energy consumption due to the complex consensus mechanisms (Fuessler et al. 2018; Neves and Prata 2018), such as proof-of-work algorithm, which requires a large amount of electricity, with one single bitcoin transaction equalling the electricity consumption of 1.5 American homes for one day (PwC 2018). There are already attempts to find solutions to mitigate high energy consumption such as using alternative consensus algorithms (e.g. proof-of-stake) (Fuessler et al. 2018; Neves and Prata 2018). The green blockchain applications, which we introduce in the following sections (also see Table 9.3), are built upon Ethereum, which requires 12–14 times less electricity than Bitcoin (PwC 2018), or upon the Stellar blockchain that requires even less electricity than Ethereum (Poseidon 2018).

Furthermore, there are several technological restrictions regarding processing time and storage of large transactional data as well as scalability (Fuessler et al. 2018). For example, due to complex verification processes transaction speed is rather low and makes blockchains more resource consuming (Citi 2016). Coupled with limited scalability this makes a widespread adoption of blockchain technology very challenging (Fuessler et al. 2018; PwC 2018). Another technological challenge is the integration of blockchain

Table 9.3 Examples of blockchain applications in green finance

Application	*Website*	*Field of application*	*Short description*	*Blockchain*	*Means of exchange*	*ICO for funding of business*
Cryptoleaf	www.cryptoleaf.io	Peer-to-peer financing and investment	Blockchain-based, tokenized equity crowdfunding platform for green projects. Funding is transferred in ETH or CLF (with lower transaction/service fees if CLF is used)	Ethereum	ETH (Ether); CLF (CryptoLEAF tokens)	Tokensale of CLF tokens
Climatecoin	www.climatecoin.io	Peer-to-peer trading platform	Peer-to-peer carbon credit trading platform where individuals can buy and trade carbon credits with emission-reducing projects via CO_2 tokens. Climatecoin purchases carbon credits from emission-reducing projects and emits CO_2 tokens respectively	Ethereum	CO_2 tokens (one token equals one carbon ton)	Tokensale of CO_2 tokens
Poseidon	www.poseidon.eco	Peer-to-peer trading platform	Platform enables retail customers to offset the carbon footprint of a purchased product via OCN tokens, which in exchange for CARBONCREDIT tokens represent the required carbon credits. Poseidon purchases carbon credits from green projects and emits CARBONCREDIT tokens respectively	Stellar	OCN tokens; CARBONCREDIT tokens (one token equals one carbon ton)	Tokensale of OCEAN (OCN) tokens
WePower	www.wepower.network	Peer-to-peer trading platform/ Peer-to-peer financing	Green energy trading platform combined with blockchain-enabled crowdfunding. Green energy providers can (e.g. to finance projects) issue their own tokens representing future energy production and directly sell it to consumers. One token represents 1 kWh of green energy	Ethereum	Energy providers' own tokens; (WPR tokens[a])	Tokensale of WPR tokens

SunContract	www.suncontract.org	Peer-to-peer trading platform	Green energy trading platform, where consumers can directly purchase energy from producers via platform's internal tokens SNC. Bid and ask price is set by consumers/producers and final price results from auction clearings on the blockchain	Ethereum	SNC tokens	Tokensale of SNC tokens
Green Assets Wallet	www.stockholmgreenfin.tech greenassetswallet.org	MRV of impact data	Platform provides investors green investment products (i.e. green bonds) including immutable validation and reporting of delivery on green impact. Platform facilitates impact reporting for issuers. Verification can be conducted by validators (e.g. certification agency)	n/a	n/a	n/a
IXO	www.ixo.foundation	MRV of impact data	IXO protocol facilitates the verification of impact data. Impact (e.g. of a project) is verified and validated through consensus mechanism (proof of impact) and stored as impact token. Flows and stocks of tokens are recorded on a public ledger	Ethereum	Impact tokens	n/a

Sources: www.climatecoin.io and whitepaper of Climatecoin (2018) (access date: 2019-01-03); www.poseidon.eco and whitepaper of Poseidon (2018) (access date: 2019-02-27); www.wepower.network and whitepaper of WePower (2018) (access date: 2019-01-03); www.suncontract.org (access date: 2019-02-25); www.stockholmgreenfin.tech and greenassetswallet.org (access date 2019-02-28); www.ixo.foundation and whitepaper of ixo foundation (2017) (access date: 2019-02-28)

This table only gives an overview of possible applications of blockchain technology in the field of climate action in Europe. The list is not comprehensive and does not raise any claim to completeness. In addition, the listed examples are all innovative businesses, mostly still in their built-up phase and need to further scale up and expand their business and underlying technology

[a]Energy providers issue their own tokens to be purchased by consumers in respective energy auctions and serving as means of exchange. In addition, WePower conducted a token sale (WPR tokens) to raise funding for its own business. WPR token holders get e.g. priority access to these energy auctions.

technology with existing systems and processes (interoperability) (Castilla-Rubio et al. 2016; PwC 2018). In addition, network and cybersecurity are seen as further risks of blockchain technology (Neves and Prata 2018).

The complexity of the blockchain technology which requires users to have a certain level of experience could further hamper a widespread adoption, since understanding builds the basis of trust and usability of an innovative technology (PwC 2018).

The emergence of innovative technologies always generates an "environment of uncertainty" (Neves and Prata 2018), which requires the adaption of existing or implementation of new legal and regulatory frameworks. For blockchain technology and its applications legal and regulatory frameworks have to be established, where the challenge lies in finding the balance between no regulation and overregulation (Neves and Prata 2018). Since innovative technologies are usually moving and changing fast, regulations have to be flexible (Neves and Prata 2018). Additionally, a blockchain is not restricted to national borders, such that legal and regulatory frameworks should be applicable and non-conflicting across jurisdictions (PwC 2018).

To enable blockchain to fully scale up and unwrap its potential in disrupting green finance, it is required to address the above-mentioned challenges with suitable measures. To this end, governments, companies and other actors have to cooperate and coordinate their actions (PwC 2018) in a quick and agile way.

Based on blockchain technology there are several applications that have a beneficial impact on the mobilization of green finance (see Fig. 9.3), which we categorize along three lines[6]: (1) peer-to-peer financing and investment, (2) peer-to-peer trading platforms and (3) measurement, reporting and verification of impact data. Beneficial features as well as limitations of blockchain apply to each field of application, accordingly.

In the following sections, we will shed light on applications of blockchain in green finance. Since the focus of this chapter is on green finance, we will elaborate the first field of application "peer-to-peer financing and investment" in more detail and will only give a brief overview on the two other fields of application.

[6] We follow the classification of UNFCCC that define four categories where blockchain can foster climate action: "improved carbon emission trading"; "facilitated clean energy trading"; "enhanced climate finance flows"; "better tracking and reporting of GHG […]" (see United Nations Climate Change (2017)).

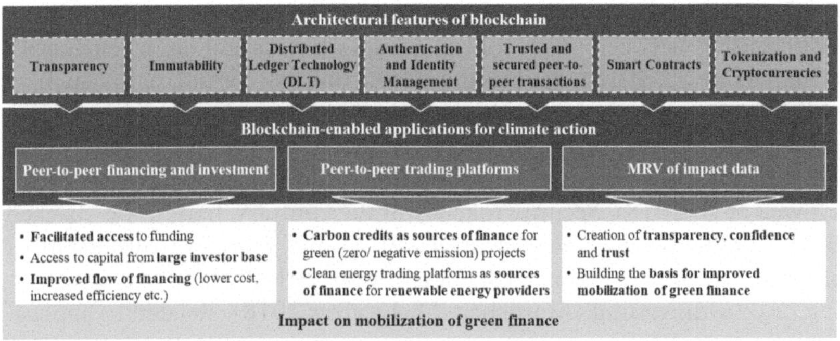

Fig. 9.3 Blockchain-enabled applications for climate action and impact on mobilization of green finance. (Sources: Authors' elaboration)

9.3.2 Peer-to-Peer Financing and Investment

We already introduced crowdfunding as one application to mobilize green finance. Blockchain also enables the raising of funds for green projects from a large number of contributors, with the key difference to "normal" crowdfunding being the means of exchange: with blockchain-enabled crowdfunding the transaction is carried out in cryptocurrency or tokens. Initial coin offerings (ICO) or token sales are a blockchain-enabled fundraising mechanism of increasing importance (Fuessler et al. 2018; PwC 2018). Especially in start-ups based on blockchain technology, ICOs are more and more replacing traditional financing (Diemers et al. 2018). This also includes green businesses that develop products or solutions—based on blockchain technology—contributing to climate action. Analogously to the specification of green crowdfunding, we will consider green ICOs as fundraising of green businesses or projects.

Definition and Functioning of ICOs

ICO is a blockchain-enabled mechanism that enables companies, start-ups or projects to raise capital (Howell et al. 2018; Li and Mann 2018) from the public by selling a "predefined number of digital tokens" (Diemers et al. 2018). Each token is a "cryptographically secured digital asset" (Howell et al. 2018) and represents a certain value or right for the owner of the token, such as the right to access or use of products or services to be produced in the future (Li and Mann 2018). During the token sale investors can acquire the project's tokens in exchange for leading cryptocurrency (e.g. Bitcoin, Ethereum) or real currency (Diemers et al. 2018).

There are different types of tokens, where a common categorization distinguishes between utility tokens, security tokens and digital currencies (Diemers et al. 2018). Utility tokens encompass certain rights such as entitling the token holder to use a future product or they can serve as a project-internal currency (Howell et al. 2018). ICOs with utility tokens are similar to reward-based crowdfunding (Amsden and Schweizer 2018; Howell et al. 2018). Security tokens, on the contrary, have to be "backed by a tangible asset" (Fuessler et al. 2018) and represent ownership or share in a company (Fisch 2018; Fuessler et al. 2018) similar to crowdlending or crowdinvesting (Amsden and Schweizer 2018). Related to applications in climate action, a security token can also represent e.g. a certain amount of CO_2 reduction (Fuessler et al. 2018). Digital currencies serve as "general-purpose medium of exchange" (Howell et al. 2018) such as established cryptocurrencies (e.g. Bitcoin or Ethereum) which can be easily exchanged for non-digital currencies (Fisch 2018). Tokenization and the use of tokens as means of exchange are also a key feature of other blockchain applications such as peer-to-peer trading and reward platforms.

Generally, token sales are executed on the webpage of the respective ventures and any transaction or interaction is directly between ventures and investors, where there are usually no dedicated platforms to host ICOs (Amsden and Schweizer 2018; Fuessler et al. 2018).

The majority of green blockchain applications presented in Table 9.3 have raised funds for their business via token sale. The tokens can then be used as means of exchange within the respective venture's economy endowing the token holder with specific rights. The blockchain-based crowdfunding platform *Cryptoleaf*[7] issued internal tokens (CLF), which token holders can use to invest in projects (listed on the platform), while benefiting from lower transaction fees compared to investing via cryptocurrency (Cryptoleaf 2018). CLF tokens also equip token holders with voting rights, regarding which projects are to be opened for funding on the platform (Cryptoleaf 2018). The green energy trading and project financing platform *WePower*[5] also raised funds via an ICO. Token holders benefit from, for example, priority access to energy auctions (WePower 2018). In addition, both platforms are similar to green crowdfunding platforms, hosting the fundraising campaigns of green projects. *Cryptoleaf* is a funding platform for mid-sized ecological projects, on which funding

[7] See Table 9.3 for an overview of applications. For more detailed information on applications see the respective webpages stated in Table 9.3.

is raised in cryptocurrencies (Cryptoleaf 2018). *WePower* enables green energy producers to raise funding via issuing their own energy tokens on the platform (WePower 2018).

Benefits

The benefits stated with crowdfunding will also hold for ICOs that additionally include beneficial features of blockchain technology. ICOs facilitate broad access to funding (PwC 2018) for green projects, start-ups and tech companies, being able to raise capital from the mass market and offer investors new investment opportunities easily accessible. Projects do not have to accomplish restrictive standards as with traditional investments and can use ICOs as a source of "rapid liquidity" (Howell et al. 2018). Blockchains' beneficial features like increased financial inclusion and access to a global network further broaden the investor base. On account of the underlying blockchain technology, investors have full transparency on all transactions (Fuessler et al. 2018). In addition, transactions can be executed directly peer-to-peer between investors and respective ventures and do not require an intermediary (Fuessler et al. 2018). In addition, without intermediaries and based on digital technology, transactions can be processed at a higher speed (PwC 2018). Hence, cost reduction and increased efficiency are also key benefits of ICOs.

Challenges and Limitations

Despite the potential, ICOs offer in facilitating funding of green initiatives there are also non-negligible drawbacks to be accounted for. ICOs do not solve the challenge of lacking standards and definitions regarding which projects can be labelled as *green*. In addition, there is still insufficient transparency and accountability regarding the information provided on the projects and the use of funds (Amsden and Schweizer 2018; BaFin 2017). Initiators of an ICO usually provide whitepapers to offer transparency on the business idea. However, information can be "objectively insufficient, incomprehensible or even misleading" (BaFin 2017), and the depth of detail, format and so on is not standardized or regulated (Amsden and Schweizer 2018). For example, verification regarding validity of the information is also insufficient due to lacking legal and regulatory guidelines (BaFin 2017). This leads to information asymmetry prevalent with ICOs involving significant risks for investors (Amsden and Schweizer 2018).

ICOs are highly speculative investments (BaFin 2017; Fisch 2018): the projects and their underlying technology are usually immature (BaFin 2017), tokens (especially utility tokens) are oftentimes lacking a tangible value since the product or service still has to be developed (Fisch 2018; Fuessler et al. 2018), prices of tokens are volatile (BaFin 2017; Fisch 2018) and liquidity in a secondary market (if existent) is not guaranteed (BaFin 2017; Howell et al. 2018). ICOs are also often linked to risks such as fraud (BaFin 2017; ESMA 2019), money laundering (ESMA 2019) and cyber insecurity (ESMA 2019). Finally, there is a lack of regulations on ICOs and crypto-assets (Fuessler et al. 2018). Additionally, jurisdictions inconsistently deal with ICOs and develop different regulatory guidelines (Diemers et al. 2018). In the EU, ICOs are allowed, however, possible regulations have to be developed (Diemers et al. 2018).

Going forward, ICOs require the development of an adequate regulation that not just prohibits ICOs (Li and Mann 2018) but is tailored to this specific "new asset class" (Chiu and Greene 2018) and that offers sufficient investor protection (Amsden and Schweizer 2018). Fuessler et al. (2018) further argue that standards and formats need to be internationally aligned.

9.3.3 Peer-to-Peer Trading and Exchange Platforms

This field includes the development of blockchain-enabled peer-to-peer trading platforms, on which individuals can securely trade clean energy or carbon credits based on tokens that represent the value of the traded good (e.g. value of quantity of energy produced) (United Nations Climate Change 2017). The trading of carbon credits can provide funding for projects with positive impact on the environment. As an example, clean energy trading platforms can facilitate green energy providers to sell future energy production to finance renewable energy projects.

Carbon Credit Trading

Carbon credits are certificates consumers can buy to offset greenhouse gas (GHG) emissions related to a purchased product or service, such as a flight ticket (Neves and Prata 2018). They function like a "compensation system" (Climatecoin 2018) in which one carbon credit equals one metric ton of GHG (e.g. carbon dioxide) (Climatecoin 2018; Neves and Prata 2018). Actors, such as companies that emit carbon dioxide buy carbon credits from other actors with zero or negative emissions, such that the emissions of the carbon credit holder are balanced out (carbon offset) (Neves and Prata 2018).

However, carbon trading is difficult to be accessed by individuals and is not directly linked to daily activities (Poseidon 2018). In addition, existing trading systems are not very transparent (Poseidon 2018) and involve transaction costs and inefficiencies like in other centralized systems. According to Neves and Prata (2018) blockchain technology and respective applications "would contribute to the functioning of the Carbon Market". First of all, via tokenization emissions can be easily traded like any asset, where one token can, for example, represent a carbon credit, which in return equals a certain amount of GHG. Tokens will all have the same value which makes trading easier compared to carbon credits that can be of different types or can have different values (Climatecoin 2018). Second, through the tokens carbon credits can be traded directly between actors emitting and those reducing GHG emissions without intermediaries and at lower cost. The blockchain further ensures that transactions are secure, immutable and transparent. Since carbon credits fund projects that have a positive impact on the environment, blockchain-enabled peer-to-peer-trading platforms for carbon credits are also a lever to mobilize green financing.

Poseidon[8] and *Climatecoin*[8] are two examples of blockchain-enabled carbon credit platforms. The platforms purchase carbon credits from selected emission reducing projects and create internal tokens, representing the respective amount of carbon credits. The tokens can then be purchased by consumers to offset their carbon footprint, serving as means of exchange (Climatecoin 2018; Poseidon 2018). In the case of *Poseidon*, retailers integrate their point of sale system and products' carbon footprint within the platform such that consumers can directly offset the carbon footprint of products at the time of purchase (Poseidon 2018). In addition, consumers can see the details of the carbon offset including the emission reducing project that is supported (Poseidon 2018).

Clean Energy Trading

Regarding the transition towards green energy, the energy market is confronted with several challenges that decelerate the broad role out of renewable energy. For example, energy trading is only between large energy producers and end users and does not facilitate peer-to-peer energy trad-

[8] See Table 9.3 for an overview of applications. For more detailed information on applications see the respective webpages stated in Table 9.3.

ing between small-scale energy producers (e.g. private persons producing solar power on their homes) and consumers (Marke 2018). Furthermore, market barriers discourage smaller companies or prosumers to enter the renewable energy market as producers (Fuessler et al. 2018; Richard and Hitchens 2018). Blockchain technology can hereby offer a solution especially regarding direct energy trading in a decentralized model (Fuessler et al. 2018). Blockchain-enabled peer-to-peer energy trading platforms allow prosumers and other small-scale energy producers to participate in the energy market without market barriers (Marke 2018) and directly sell energy to consumers at lower transaction costs (Fuessler et al. 2018; Marke 2018). Based on the blockchain features, energy transactions are secure, transparent and verifiable (Marke 2018). The tokenization of renewable energy as tradable asset can further foster investments in renewable energy (Fuessler et al. 2018).

Platforms, such as *WePower*[9] or *SunContract*,[9] enable the peer-to-peer trading of energy via tokens which serve as medium of exchange. Consumers can either get the tokens via token sale or purchasing them on an exchange, where each token represents a certain amount of energy and gives the token holder the right to receive this amount of energy. The producer receives tokens in return for the energy produced and can then exchange the tokens for other cryptocurrency or real currency. Final price and quantity of energy is set via auctions on the blockchain (SunContract 2017; WePower 2018). Consumers hereby benefit from lower energy cost and choosing renewable energy whereby producers benefit from higher compensation due to the direct sale to consumers (SunContract 2017).

On the platform *SunContract* energy providers and consumers can directly trade electricity via *SunContract's* internal SNC tokens (SunContract 2017), whereas on the platform *WePower* energy providers issue their own tokens, in order to sell future energy production (WePower 2018). Consumers can purchase this future energy upfront below market rates via purchasing a provider's internal tokens, which entail the right to use the energy when it is produced or trade the energy on the platform (WePower 2018).

[9] See Table 9.3 for an overview of applications. For more detailed information on applications see the respective webpages stated in Table 9.3.

9.3.4 Measurement, Reporting and Verification of Impact Data

An important lever to mobilize green finance is the possibility for investors to have transparency on the flow of finance and sustainability of the investment target, such as the reporting of a green bond issuer on the use of proceeds and the positive environmental impact of the underlying project. To quantify this impact, respective impact data has to be collected, measured as well as verified (e.g. the level of reduction in CO_2 emissions the funded project has achieved). The principle of MRV was hereby introduced to create transparency regarding the impact of climate action (Fuessler et al. 2018). However, according to Fuessler et al. (2018) there are five key challenges involved with MRV which are (1) "lack of trust in data", (2) "costly, complex collection of data", (3) "costly and complex impact quantification and reporting", (4) "quality assurance and quality control" and (5) "costly verification of emission reductions". Impact reporting lacks standardization, consistency and comparability, involves significant resources (Climate Bond Initiative 2017) and underlying data is often rather weak, based on estimations (Sanderson 2018). Sanderson (2018) further underlines the importance of transparent and secure verification and monitoring of impact data, building the basis of standards and definitions or regulation in general.

Blockchain can overcome these challenges with MRV, can improve the overall MRV process (increased efficiency, lower costs (Fuessler et al. 2018)) and can facilitate the creation of transparent, trusted and verified impact data. Any transaction or data can be traced back to its origin (Fuessler et al. 2018) and is always verified by the network. The collection of impact data is more efficient and less costly since blockchain and digitization allow for automation of processes at higher speed and accuracy (Fuessler et al. 2018). In addition, data can be securely and immutably stored on the blockchain (Meunier 2018). Automation and smart contracts will also improve impact reporting and quality assurance (Fuessler et al. 2018). Finally, blockchain technology in combination with other digital technologies (e.g. AI), makes the verification of emission reduction more efficient (Fuessler et al. 2018) and can avoid the risk of double-counting (Sanderson 2018).

For example, the *IXO Foundation*[10] created the *ixo protocol*, a "decentralized impact evaluation protocol", within which impact claims from ser-

[10] See Table 9.3 for an overview of applications. For more detailed information on applications please refer to the respective webpages stated in Table 9.3.

vice providers are verified by an evaluator (e.g. individuals, software algorithms) and after being qualified as "Proof of Impact" stored as "Impact Tokens" to a public ledger (IXO foundation 2017). Both service providers and funders are hereby provided transparency on the true impact of initiatives, and the risk of fraud and double-counting is reduced (IXO foundation 2017; Meunier 2018).

Efficient MRV and high-quality trusted and secure impact data is also important regarding energy trading and carbon emissions trading. For example, in the case of carbon credits a prerequisite is the tracking of the carbon footprint of a product in order to calculate carbon credits required to offset the carbon footprint as well as the quantification of the impact of a negative emission project in order to create the respective amount of carbon credits.

9.3.5 *Overview on Real-World Examples of Blockchain Applications in Green Finance*

Table 9.3 gives an overview on existing blockchain applications in green finance in Europe for each of the three fields of application. The examples listed in Table 9.3 show how blockchain technology can promote the mobilization of finance towards climate action (also see Figs. 9.2 and 9.3).

First of all, most applications have raised funds via ICO. Since they can all be classified as green businesses (being beneficial to climate action), each ICO enabled the mobilization of green finance. Second, peer-to-peer energy trading platforms such as *WePower* and *SunContract* lower the barriers for small green energy providers to enter the energy market by enabling them to directly trade with consumers, which results in a better compensation. Hence, peer-to-peer energy trading platforms can encourage energy providers and mobilize respective investments into renewable energy in the first place. *WePower* in addition, enables green energy providers to raise funding for their renewable energy projects. Third, peer-to-peer carbon credit platforms such as *Climatecoin* and *Poseidon* facilitate the access to carbon credit markets and respective trading of carbon credits for any individual and hereby foster the direct flow of financing towards emission reducing projects. Finally, the prerequisite of these blockchain applications and overall mobilization of green finance—measurement, reporting and verification of high-quality impact data—is provided by applications like *Green Assets Wallet* and *IXO Foundation (ixo protocol)*.

9.4 Conclusions

The objective of this chapter is to delineate the potential of fintech and blockchain with respect to mobilizing green finance by explaining the key functionalities of respective applications including their key benefits and limitations. Moreover, we illustrate fintech and blockchain applications in green finance on the basis of example cases, introducing companies or projects in Europe that contribute to this mobilization using these innovative technologies. This demonstrates that fintech and blockchain applications facilitate access to funding for green initiatives, they can enlarge the investor base including small investors and private capital and provide new forms or mechanisms of financing. In addition, they operate in a decentralized system avoiding intermediaries, decreasing costs and inefficiencies. Finally, in particular the applications using blockchain technology allow for increased transparency and accountability reducing the risk of greenwashing. Hence, fintech and blockchain (in combination with other digital technologies) have the capability to overcome the barriers of mobilizing green finance.

However, future success and a widespread application of fintech and blockchain in green finance is conditioned on the trust and confidence of the users, the usability and applicability of these technologies as well as on adequate legal and regulatory frameworks. Innovative applications will only become effective and work efficiently if a sufficient number of market players uses them. This is preconditioned upon users perceiving these technologies as applicable, comprehensible, trustworthy and secure, which in turn requires agile regulatory and legal frameworks to build upon despite the fast-changing environment. In addition, uniform standards and definitions for green finance are required to create trusted labels and certificates in order to incentivize investors for green investing as well as to empower the recipients of green financing to prove their greenness. Another requirement is the technical feasibility, implying that current technological risks and flaws are solved and the public literacy with respect to the innovative technologies is increased. In order to unlock the full potential of fintech, blockchain and other digital technologies in green finance, governments, the financial sector, companies and the technological sector need to cooperate cross-nationally to provide uniform definitions and standards, required regulatory and legal frameworks. Furthermore, it appears necessary to constantly drive technological advancement and innovation.

References

Adhami, S., Giudici, G., & Anh, H. P. N. (2017). *Crowdfunding for green projects in Europe: Success factors and effects on the local environmental performance and wellbeing.* Retrieved from http://www.crowdfundres.eu/wp-content/uploads/2017/11/Crowdfunding-for-green-projects-in-Europe-2017.pdf

Alonso, P., Caneva, S., Maidonis, T., et al. (2017). Policy recommendations on regulatory and market framework improvements for crowdfunding RES projects. *CrowdFundRES.* Retrieved from http://www.crowdfundres.eu/wp-content/uploads/2017/07/CrowdFundRES_-Policy-Recommendations.pdf

Amsden, R., & Schweizer, D. (2018). Are blockchain crowdsales the new "gold rush"? Success determinants of initial coin offerings. *SSRN Electronic Journal.* https://doi.org/10.2139/ssrn.3163849.

BaFin. (2017). *BaFin – Consumer warning: The risks of initial coin offerings.* Retrieved February 19, 2019, from https://www.bafin.de/SharedDocs/Veroeffentlichungen/EN/Meldung/2017/meldung_171109_ICOs_en.html

Beioley, K. (2019). Robo advisers short circuit over ethical investing. *Financial Times.* Retrieved from https://www.ft.com/content/1dabbce8-b5d8-11e8-b3ef-799c8613f4a1

Belleflamme, P., Lambert, T., & Schwienbacher, A. (2014). Crowdfunding: Tapping the right crowd. *Journal of Business Venturing, 29*(5), 585–609. https://doi.org/10.1016/j.jbusvent.2013.07.003.

Bergmann, A., Betz, S., Burton, B., Kohl, K., Maidonis, T., & Klaes, M. (2016). Crowdfunding for renewable energy: Survey results on public perceptions and the views of crowdfunding platforms and project developers. https://doi.org/10.13140/RG.2.1.2580.8888.

Blakstad, S., & Allen, R. (2018). International sustainable investment. In *Fintech revolution: Universal inclusion in the new financial ecosystem* (pp. 215–226). Cham: Springer. https://doi.org/10.1007/978-3-319-76014-8_13.

Bonzanini, D., Giudici, G., & Patrucco, A. (2016). The crowdfunding of renewable energy projects. In *Handbook of environmental and sustainable finance* (pp. 429–444). Elsevier. https://doi.org/10.1016/B978-0-12-803615-0.00021-2.

Bradford, C. S. (2012). Crowdfunding and the federal securities laws. *2012 Columbia Business Law Review, 1*(1), 1–150.

Butticè, V., Colombo, M. G., Fumagalli, E., & Orsenigo, C. (2019). Green oriented crowdfunding campaigns: Their characteristics and diffusion in different institutional settings. *Technological Forecasting and Social Change, 141*, 85–97. https://doi.org/10.1016/j.techfore.2018.07.047.

Caneva, S., & Alonso, P. (2018). *Crowdfunding Renewable Energy – A practical guide for Crowdfunding Platforms.* Project Developers, Investors and Policy Makers. Retrieved from http://www.crowdfundres.eu/index.html@p=1191.html

Castilla-Rubio, J. C., Zadek, S., & Robins, N. (2016). Fintech and sustainable development: Assessing the implications. *UNEP Inquiry*. Retrieved from http://unepinquiry.org/wp-content/uploads/2016/12/Fintech_and_Sustainable_Development_Assessing_the_Implications.pdf

Chiu, I. H.-Y., & Greene, E. (2018). Can ICO markets offer insights into marketising sustainable or social finance? Proposals for a revolutionary regulatory framework. *SSRN Electronic Journal*. https://doi.org/10.2139/ssrn.3290762.

Citi. (2016). *Digital disruption – How fintech is forcing banking to a tipping point*. Retrieved from https://ir.citi.com/D%2F5GCKN6uoSvhbvCmUDS05SYsRaDvAykPjb5subGr7f1JMe8w2oX1bqpFm6RdjSRSpGzSaXhyXY%3D

Climate Bond Initiative. (2017). *Post-issuance reporting in the green bond market*. Retrieved from https://www.climatebonds.net/files/files/ClimateBonds_Green_Bonds_Post_Issuance_Use_of_Proceeds_Report_270617%284%29.pdf

Climatecoin. (2018). *Whitepaper – Climatecoin*. Retrieved from www.climatecoin.io

CRIC. (2018). *Fintech action plan der EU-Kommission: Nachhaltigkeit fehlt*. Retrieved March 4, 2019, from https://www.cric-online.org/info-medien/news/363-fintech-action-plan-der-eu-kommission-kommt-ohne-nachhaltigkeit-aus

Cryptoleaf. (2018). Whitepaper – Cryptoleaf v2.8. Retrieved from https://www.cryptoleaf.io/

Diemers, D., Arslanian, H., McNamara, G., Dobrauz, G., & Wohlgemuth, L. (2018). Initial Coin Offerings: A strategic perspective. *PwC Strategy &*. Retrieved from https://cryptovalley.swiss/wp-content/uploads/20180628_PwC-S-CVA-ICO-Report_EN.pdf

Dilger, M. G., Jovanović, T., & Voigt, K.-I. (2017). Upcrowding energy co-operatives – Evaluating the potential of crowdfunding for business model innovation of energy co-operatives. *Journal of Environmental Management, 198*, 50–62. https://doi.org/10.1016/j.jenvman.2017.04.025.

Dorfleitner, G., Hornuf, L., Schmitt, M., & Weber, M. (2017). The Fintech market in Germany. In *Fintech in Germany* (pp. 13–46). Cham: Springer.

ESMA. (2019). *Advice – Initial coin offerings and crypto-assets*. Retrieved from https://www.esma.europa.eu/sites/default/files/library/esma50-157-1391_crypto_advice.pdf

European Commission. (2018a). *Action plan: Financing sustainable growth*. Retrieved from https://eur-lex.europa.eu/legal-content/EN/TXT/PDF/?uri=CELEX:52018DC0097&from=EN

European Commission. (2018b). *Fintech action plan: For a more competitive and innovative European financial sector*. Retrieved from https://eur-lex.europa.eu/resource.html?uri=cellar:6793c578-22e6-11e8-ac73-01aa75ed71a1.0001.02/DOC_1&format=PDF

Fein, M. L. (2015). Robo-advisors: A closer look. *SSRN Electronic Journal*. https://doi.org/10.2139/ssrn.2658701.

Fisch, C. (2018). Initial coin offerings (ICOs) to finance new ventures: An exploratory study. *Journal of Business Venturing, 34*(1), 1–22.

Fisch, J. E., Labouré, M., & Turner, J. A. (2018). *The emergence of the robo-advisor.* Retrieved from https://repository.upenn.edu/cgi/viewcontent.cgi?article=1010&context=prc_papers

Fromm, K. F. (2018). *Wie funktioniert der Anlagestil LIQID Global Impact?* Retrieved February 11, 2019, from https://www.liqid.de/de/magazin/investieren/wie-funktioniert-der-anlagestil-liqid-global-impact

Fuessler, J., León, F., Mock, R., et al. (2018). *Navigating blockchain and climate action.* Retrieved from https://www.goldstandard.org/sites/default/files/documents/cli_report_dec18.pdf

Galen, D., Brand, N., Boucherle, L., et al. (2018). *Blockchain for social impact – Moving beyond the hype.* Retrieved from https://www.gsb.stanford.edu/sites/gsb/files/publication-pdf/study-blockchain-impact-moving-beyond-hype.pdf

Global Sustainable Investment Alliance (GSIA). (2016). *2016 Global Sustainable Investment Review.* Retrieved from http://www.gsi-alliance.org/wp-content/uploads/2017/03/GSIR_Review2016.F.pdf

Howell, S., Niessner, M., & Yermack, D. (2018). Initial coin offerings: Financing growth with cryptocurrency token sales. *SSRN Electronic Journal.* https://doi.org/10.2139/ssrn.3201259.

IXO foundation. (2017). *Whitepaper – ixo foundation.* Retrieved from https://ixo.foundation/

Kaya, O. (2017). *Robo-advice – A true innovation in asset management.* Deutsche Bank Research. Retrieved from https://www.dbresearch.com/PROD/RPS_EN-PROD/PROD0000000000449125/Robo-advice_-_a_true_innovation_in_asset_managemen.PDF

Lam, P. T. I., & Law, A. O. K. (2016). Crowdfunding for renewable and sustainable energy projects: An exploratory case study approach. *Renewable and Sustainable Energy Reviews, 60,* 11–20. https://doi.org/10.1016/j.rser.2016.01.046.

Leblanc, R. (2017). *Crowdfunding websites and best practices for green businesses.* Retrieved from https://www.thebalancesmb.com/green-business-crowdfunding-guide-4135794

Li, J., & Mann, W. (2018). *Initial coin offering and platform building.* Retrieved from https://pdfs.semanticscholar.org/309e/f98741d5da2003df8317fd605e1ac83d6fb9.pdf

Marke, A. (2018). *Transforming climate finance and green investment with blockchains.* Academic Press. https://doi.org/10.1016/C2017-0-01389-7.

Meunier, S. (2018). Blockchain 101: What is blockchain and how does this revolutionary technology work? In *Transforming climate finance and green investment with Blockchains* (pp. 23–34). Academic Press. https://doi.org/10.1016/B978-0-12-814447-3.00003-3.

Neves, L. P., & Prata, G. A. (2018). *Blockchain contributions for the climate finance: Introducing a debate.* FGV, International Intelligence Unit. Retrieved from https://www.kas.de/c/document_library/get_file?uuid=ea6109a2-7677-9bfa-d4d0-6cbae35ebcc7&groupId=252038

O'Keefe, D., Wamund, J., & Lewis, B. (2016). *Robo advising – Catching up and getting ahead*. Retrieved from https://www.roboadvisor-portal.com/wp-content/uploads/2018/01/Robo-Advising-Catching-Up-And-Getting-Ahead.pdf

Plunkett, E., Sabhlok, V., & Acharya, M. (2016). *Mind the gap – Bridging the climate financing gap with innovative financial mechanisms (Insight Br)*. Retrieved from http://gggi.org/site/assets/uploads/2016/11/2016-11-Mind-the-Gap.pdf

Poseidon. (2018). *Whitepaper – Poseidon*. Retrieved from www.poseidon.eco

PwC. (2018). *Building block(chain)s for a better planet*. Retrieved from https://www.pwc.com/gx/en/sustainability/assets/blockchain-for-a-better-planet.pdf

Ramos, S., Laurent, P., Hugé, F.-K., & Duprel, C. (2016). *How can Fintech facilitate fund distribution?* Retrieved from https://www2.deloitte.com/content/dam/Deloitte/lu/Documents/technology/lu_how-can-fintech-facilitate-fund-distribution.pdf

Richard, A., & Hitchens, R. (2018). Efficient power markets: Reimagining the global market with ethereum. In *Transforming climate finance and green investment with blockchains* (pp. 93–98). Academic Press. https://doi.org/10.1016/B978-0-12-814447-3.00007-0.

Sanderson, O. (2018). How to trust green bonds: Blockchain, climate, and the institutional bond markets. In *Transforming climate finance and green investment with blockchains* (pp. 273–288). Academic Press. https://doi.org/10.1016/B978-0-12-814447-3.00020-3.

Shishlov, I., Morel, R., & Cochran, I. (2016). *Beyond transparency: Unlocking the full potential of green bonds*. Retrieved from https://www.i4ce.org/wp-core/wp-content/uploads/2016/06/I4CE_Green_Bonds-1.pdf

Statista.com. (2019). *Robo-advisors – Europe | Statista market forecast*. Retrieved March 8, 2019, from https://www.statista.com/outlook/337/102/robo-advisors/europe

SunContract. (2017). *Whitepaper – SunContract*. Retrieved from https://suncontract.org/tokensale/res/whitepaper.pdf

United Nations Climate Change. (2017). *How blockchain technology could boost climate action*. Retrieved February 13, 2019, from https://unfccc.int/news/how-blockchain-technology-could-boost-climate-action

Vasileiadou, E., Huijben, J. C. C. M., & Raven, R. (2016). Three is a crowd? Exploring the potential of crowdfunding for renewable energy in the Netherlands. *Journal of Cleaner Production, 128*, 142–155. https://doi.org/10.1016/j.jclepro.2015.06.028.

WePower. (2018). *Whitepaper – WePower v0.8*. Retrieved from https://wepower.network/media/WhitePaper-WePower_v_0.81-1.pdf

CHAPTER 10

Sustainable Finance: A Common Ground for the Future in Europe?

Silvio Goglio and Ivana Catturani

10.1 Introduction

Sustainable finance is a relatively recent term. Its diffusion is mainly due to a reaction to two circumstances not directly connected: the distortion of the role of the financial system and the environmental emergency. Consequently, it contains and somehow integrates two meanings.

Since the causes of the financial crisis that broke out in 2007 have become evident, the activity of the financial sector is not only under scrutiny of domestic and international regulators but also under the lens of a wider public including clients, investors, employees and stakeholders in general. In the two decades before the crisis, finance had—and largely still has—lost sight of its instrumental nature, being increasingly considered as an end in itself. Alongside the disproportionate increase in salary and profits, technological development and globalization have enhanced the growth in the size of the sector, largely engaged in activities poorly connected to the real economy and dominated by speculative reasons (Silver

S. Goglio (✉) • I. Catturani
University of Trento, Trento, Italy
e-mail: silvio.goglio@unitn.it; ivana.catturani@unitn.it

M. Migliorelli, P. Dessertine (eds.), *The Rise of Green Finance in Europe*, Palgrave Studies in Impact Finance,
https://doi.org/10.1007/978-3-030-22510-0_10

2017). As a result, the public opinion is distancing by seeing finance no longer as a tool of growth but as a form of appropriation of income and wealth. At the same time, global pollution, climate change and the unsustainable exploitation of resources, beyond their capacity of reproducing, are increasingly palpable and worrying issues.

By sustainable finance, we can therefore identify both a non-predatory finance, more attentive to the production, than to the extraction, of value, and a finance aimed at fostering development sustainable in the long term. The two meanings can obviously be integrated into one that includes both. A stronger targeted financial system is essential for achieving a successful transition to a new pattern of development more responsible and inclusive: the environmental dimension must integrate into a more complex frame designed to increase the value generated by investments. The aim is not only to guarantee resources but also to implement social objectives, generating at the same time an economic return for investors. This involves the integration of environmental, social and governance features with the mission of harnessing resilience, targeting capital allocation and improving accountability. It is a multidimensional approach that deals with ethical questions, environmental and climate issues, social responsibility considerations and risk management requirements.

To somehow redeem finance, its concept has been related with ethical and sustainable attributes, such as "social finance", "impact finance", "ethical banking" and "social and solidarity finance". It is not just a terminological issue, but the choice of a new way to address resources for specific goals. The commitment by the financial world to take concrete actions to curb climate change and enhance the life condition of vulnerable people can be seen as the apex of a decade of effort to distance from the speculations that led to the crisis. The role that finance can play in addressing social and environmental needs is pivotal. The structural contraction of public resources and the social changes open the arena for new actors to drive the search of innovative tools. As a result, a closer cooperation between finance and social and environmental dimensions arises, which can create great potentialities to support the modernization of social and economic development policies. For finance this is a great opportunity to prove its capacity to add value in the economy.

This view is asserted by the 2015 Sustainable Development Goals and the Paris Agreement: the agenda they generated is intense and challenging. However, how underlined by Dombret (2018) in his speech at the 20th Bundesbank symposium "Banking supervision in dialogue":

> If the global community is even half-way committed to hitting the ambitious target of 2 degrees Celsius, there will need to be some far-reaching changes to the economic systems as we know them. And as for the time frame, the later we get started, the deeper our intervention will have to be. [...] Every area of the economy will need to adjust by correctly pricing the externalities of climate change and internalising them. And those adjustments mean more than simply trimming our CO_2 emissions. They will transform the entire way in which we do business and affect the path along which the economy and society are progressing.

The changes introduced here are more than a new combination of product features to capture a larger share of market. The challenge for the economy, in general, and for the financial sector, in particular, is to introduce a new paradigm in which the creation of value is at the core. Sustainable finance is a financial world which looks primarily at the long-term repercussions of its actions (Dombret 2018). In this process, the entire sector is required to intervene, not only cooperative or ethical banks.

Sustainability concerns the challenges and risks of failure together with opportunities. The core of any sustainability approach is the awareness of the relevant impact areas and the definition of the appropriate risk management strategies. There is a shared recognition that the so-called residual factors, as social capital and institutional framework, for a long time considered marginal, are able to create value.[1] As much as the financial system is progressively taking into consideration environmental and social factors in the allocation of its resources, the promising development in this contest has not yet achieved a systematic impact across the financial mainstream due to (UN Environment Inquiry 2016)[2]:

[1] The UN Environment Inquiry (2016) identifies only three factors: environment, society and governance. However, according to the authors of this chapter, this view is reductive since more elements, such as immaterial factors, might contribute in the process (Goglio 2002).

[2] The UN Environment Inquiry report on Italy includes in this list the limited access to finance, especially for SMEs, since it reduces their participation into green economy. In our view, SMEs might get advantages in investing in this sector given its potentiality. However, in their case, the limited access to finance is not a stringent limit, since their main sources of financing are the internal mechanism, such as collecting capital by the owners or angel finance.

- Unpriced environmental externalities that can tilt the risk/return profile away from sustainable finance.
- Financial decision-making who does not adequately take into consideration the long-term challenge of these investments.
- The achievement of sustainable strategy, which might result only in a reputation enhancement for the supplier.

Financial capital is not therefore a goal in itself, but a vehicle to achieve other goals, ultimately to produce a blended output of goods and services, for consumption and investment, which benefits the society. This means that financing should be addressed to economic activities able to generate social and environmental benefits, and financial profit has its raison d'être in the capacity to convey resources into such activities. However, the question is: which goods, services and investments really benefit the society and the environment and how can we measure its usefulness? The matter is that usefulness might be differently perceived from subject to subject. Indeed, the concept of utility developed by the marginalist approach is subjective: when the subject's needs (even the externally induced ones) are satisfied, though in contrast with the wellbeing or even the survival of the system, we have production of utility. Utility may be therefore in contradiction with sustainability. This leads us to consider the meaning of value and the processes of its creation and appropriation.

10.2 The Need for a Value-Added Approach to Sustainable Finance

The term *value* refers to the process of generating a surplus through the production of both tangible and intangible outcomes: the concept of wealth can be a synonymous, depending on how it is calculated. The process of distribution and redistribution of income through the price system can lead to the extraction of part of the value from the producers to subjects not involved in the production of the same (the rentiers): rent requires economic and/or political power, ultimately a monopolistic position. While the distinction between these two concepts—that is, value and rent—was crucial to the classic economists, it has become less relevant in the new schools of thought, up to the point of considering the extraction itself of value as productive and, as a consequence, to be included in the GDP computation. Following Mazzucato (2018), we maintain this distinction in our theoretical frame, trying to measure the role of the financial system in the creation of a value-based economy.

The way in which value and rent are identified influences the evaluation of financial system role. This can be clearer if we consider how the concept of rent has changed in economic thoughts. From an income originated by a non-productive activity, the rent has been seen as a reward for the marginal productivity of capital and land, likewise the determination of profit. Both the classic and the neoclassic approach see the rent as a monopolistic income; however, its nature is intrinsically different in the two schools because of the different theories of value at their base. For the classics, the rent is an income derived from the control of scarce resources not produced. In the neoclassical frame, since income must match productivity, there is no room for rent, understood as a gain in exchange for nothing. Marshall relaxes this result, including the quasi-rent, which differs from pure economic rent in that it is a temporary phenomenon. It can arise from the barriers to entry that potential competitors face in the short run, such as the granting of patents or other legal protections for intellectual property by governments. It can also emerge due to the entrepreneurial address of market fluctuation, or it can arise due to the lack of real capital to meet near term demand increases. In the longer term the opportunity to profit will bring new capital into existence and the quasi-rent will be competed away. Unearned income, seen by Smith and Ricardo as a parasitic behaviour, is considered in the mainstream economy just as an impediment to perfect competition equilibrium (Mazzucato 2018).

As we said, value is intended as a surplus, generated by the production of goods and services, net of direct and indirect costs. Once created at the micro level, value can be aggregated at the macro level in order to calculate GDP. About this, we need to clarify some points.

The first point refers to the production process. The ultimate factor of production is the knowledge embedded in capital and in labour (Marshall 1920, 4, I); capital may be physical, human and social. Land is a factor of production once its fertility is used to grow products. Natural resources are involved in the production as inputs but are not factors since they are not able to produce per se. Their use can be either sustainable or not sustainable in the long term. In our frame, a renewable resource is sustainably utilized when it is employed at a rate lower than its capacity of regenerate itself, or it can be substituted by other inputs. Also non-renewable resources can be substituted, thanks to progress in knowledge. In these cases, production is not affecting the sustainability of the ecosystem. When resources are used in an unsustainable way—that is, when they are consumed at a rate higher or equal to their capacity to replicate themselves or valid substitutes are not introduced—this affects the creation of value as a negative externality.

Positive and negative externalities created in the process of production make that the value created might actually be higher or lower if compared to that deductible using the market prices system. In particular, negative externalities should be included in the national accountability with a negative sign, while positive externalities increase the net value. Statistical methods taking into account the externalities impact, if included in the national accountability, might describe the economy in a more precise way: indeed, an index calculating net social value could be more appropriate than current gross domestic product to compare economies.[3] Time horizon is relevant in the internalization of externalities, since the resultant net value might change if short or long term is considered. Positive externalities can require more time to be realized and evaluated, in particular, when they act as promoters of social and immaterial factors, as the level of civil engagement and recognition or ecological awareness.

A second point to clarify is the "detection of value". The productive process might indeed give rise both to value and to non-value, creating non-value when it exploits resources at a non-sustainable rate. It is not just a matter of balancing positive and negative externalities. As remembered above, production is characterized by externalities, positive or negative, independent of the way in which resources are employed. In particular, negative externalities are present even when the creation of goods and services is sustainable. On the contrary, when the use of resources is at a rate higher than their replacement, the result is non-value. Once again, non-value should be included in the national accountability with a negative sign.

A third point refers to the distribution of value and to the capacity and power of rent seeking groups to expropriate part of it. The unproductive process of rent seeking redistributes the net output to actors able to exploit a monopolistic privileged position. The result is an unfair redistribution,

[3] Many international organizations have introduced indices aimed to compare economic systems not on the mere GDP, but adding other relevant aspects, such as health and education (see the Human Development Index by the UNDP). What we propose here is not to add more items to the traditional GDP but to change the way in which the product is accounted by "cleaning" the value from negative externalities while adding positive ones.

where the unproductive activities grow at the expense of the productive business. The main negative aspects are that the rent-seeking activity is less interested in reinvesting in productive process, while more engaged to maintain its position, paving the road to a less efficient system in the long term.

To sum up, in an approach based on the theory of value, production generates value for the economy in the form of surplus, employing knowledge as a factor of production embedded in capital, labour and social relationships; natural resources take part as inputs. As by-products, positive and negative externalities are generated and should be accounted with respectively a positive and a negative sign to describe the actual outcomes of the economy. However, should the productive process exploit resources in a non-sustainable way, the result is non-value. The next question is what is the role of finance in this frame?

10.3 The Role of Value-Based Finance

The severity of the financial crisis in 2007 has pressed the attention of both the economists and the public opinion on the topic of the separation between the creation and the extraction of value. As underlined by Mazzucato (2018), before the emergence of the financial crisis, the income share of the richest 1% of the US population grew from 9.4% in 1980 to 22.6% in 2007. To generate gains without producing a surplus, but simply asking for prices higher than the competitive market ones, and cutting out competitors, is the way followed by the so-called *takers* to increase their income at the expense of the *makers*, who, on the contrary, create value. And financial intermediaries often fall into the group of takers.

The main allegation to banks and financial institution after the financial turmoil has been to extract profits from speculative transactions without adding value to the economy, by imposing an unjustified spread between buying and selling prices. The productive world of the factories has been contraposed to the rent-taking financial sector, opposing the "good guy"—the real economy—to the "bad guy"—the financial economy. However, such division of the world is too simplistic, since financial services do also play a crucial role in market interactions and investments. The question is how to shape these activities so that they can be instrumental to the production of value, supporting correct and sustainable use of the factors of production and of natural resources.

We cannot point out financial intermediaries as evil regardless of their way of operating. We can recognize value-based financial intermediaries when:

- They drive capitals in activities intensive in green and social productive capital.
- They support firms whose production is aimed at preserving the environment and at including vulnerable subjects.
- There is a human design in the capital invested.
- They exploit factors for a sustainable production.

In some case, financial intermediaries are considered "values"-based actors as long as they invest in projects aimed at enhancing vulnerable people or at saving the natural environment. The added label "values" implies an ethical understanding of the role of the intermediary. However, the mere fact of selling products, which supply capital for green or social projects, is not enough to include the financial intermediary in the list of the "good". As an example, the diffusion of green bonds, given the appealing of the market, could not be addressed at the creation of value, but it might be a strategy to extract rent from a speculative and growing market (see below). Moreover, banks can include some values-driven products to improve their reputation. Therefore, the simple involvement of financial products in the social and environmental sphere, though it is sufficient to bear the "values" brand, might not be enough to enhance the creation of actual value in the economy.

According to this view, to evaluate the coexistence of profit-driven and value and sustainable-based goals, it is important to analyse the real objective of the financial actors. Their institutional nature is not sufficient to detach their will to create or to extract value, that is, it is not guaranteed that a not-for-profit bank will avoid a speculative behaviour, while a profit-driven bank is not genuinely interested in supporting a sustainable project. To better understand this dichotomy, it is necessary to distinguish the meaning of the word *sustainability* when applied to intermediaries and when applied to financial products.

10.3.1 *Financial Actors and Sustainability*

Banks are not the only supplier of financial products dedicated to sustainability, in general, and to the green finance, in particular. On the one hand, the public sector and the regional institutions are active in proposing financial solutions with the main objective of enhancing the achievement of the Sustainable Development Goals. On the other hand, non-intermediated forms of financing are emerging, thanks to the diffusion of the online instruments: crowd funding, for instance, allows the disintermediation of credit using mainly online platforms and can provide many types of financing tools, such as equity, loans, prizes or donations. However, the focus here is mainly on financial intermediaries, given their structured role in the economy and their business model.

We suggest a classification that goes beyond the usual institutional one, based on the actual capacity of the bank to support the social and green economy. In particular, we can distinguish among three types of banks committed towards sustainability not only through the selling of sustainable products:

- Profit-driven banks sustaining direct costs to support sustainable projects (reducing their earnings in terms of interest rates).
- Not-for-profit banks ready to diversify their traditional business towards sustainable projects (possible less profitable and more difficult to justify to their members).
- Specialized banks fully committed towards sustainability.

Three main dimensions should be considered when analysing the compliance of banks' operations with the sustainability requirements, that is, the business, the social and environmental, and the governance dimensions.

The Business Dimension

We will consider a bank as a value-based actor whether it is able to generate a surplus by its activity. The surplus should be intended as in the above theoretical frame—that is, as the difference between value and costs and not as rent. Banks mainly base their gains on both spreads between active and passive interest rates, and intermediation. Their productivity has been measured according to their capacity on these two margins. Cooperative banks and specialized banks, in particular, show a business model in which the main source of earnings remains the interest rate spread. This is partially justified by the risk assumed by the intermediary in its lending activity. However, when this difference is excessively high, the bank might hide a rent extraction strategy. In this case, even banks with a not-for-profit institutional form cannot be considered as value-based financial institutions. On the contrary, banks that decide to invest their capital in value-driven activities and not in speculative business should be regarded as value-based intermediaries.

The Social and Environmental Dimension

The social and environmental goals of a sustainable bank, even though not specifically expressed, should aim to positively impact the communities and the environment they serve: achieving this goal requires a long-term perspective. Moreover, it is pivotal to include a set of stakeholders larger than the mere group of shareholders. The sustainability of the financial intermediary is constituted by a direct and an indirect impact. The direct aspect refers, for example, the choice of reducing the use of paper or the consumption of energy or the attention towards gender discrimination in carrying out the daily activities of the bank. The indirect part refers to the sustainable impact of the projects financed. While it is easier for local banks, traditionally more integrated into the social and economic network of the served area to show its sustainable efforts, for larger banks it is more challenging. However, the diffused network of branches, the employment policies and the capacity of learning about clients' behaviours through internet devices may cut the distance between the profit-driven (and larger size) bank and the community of interest. In this logic, in financing projects banks may play an educational role by monitoring and by pushing firms to operate in an eco-friendly and a socially responsible way.

The Governance Dimension

The inclusive governance of sustainable intermediaries implies participatory and often democratic decisional processes. Targets are fixed by involving a significant number of stakeholders such as employees, owner, final users and the local community. The democratic voting system, distinctive of cooperative banks, does not guarantee per se a higher level of participation into the governance, since it offers few incentives for investors to own more share and thus to participate actively in the management and control of the bank. Other mechanisms, which enhance the bottom-up approach and the inclusion of stakeholders, might be implemented by non-cooperative banks, with good results in terms of participation, such as open forums to collect clients' wishes or mobile applications to increase the participation of the youngest in the banking activities.

10.3.2 Sustainable Finance Products

While it is a Copernican revolution to shift from a traditional to a sustainable form of business, offering sustainable products can be just considered as a market strategy. As a matter of fact, such products are offered on the market by non-intermediary actors (such as the online platforms), whose sustainable features are not indeed determinant for the consumer's choice. To be qualified as sustainable, products must provide the consumer a transparent option to reduce the indirect impacts of their banking activities, that is, to decrease significantly negative environmental impacts or provide social benefits. Sustainable investment products represent a niche yet promising market. Their share in 2018 was about 5–10% (depending on the scope and definition used), while sustainable savings products had a share of about 1% of total savings (Sustainable Saving and Investing 2019). To have any meaningful impact on the transition towards a sustainable economy and society, the supply of these products should go beyond this niche and reach the mainstream financial products and services.

There are several kinds of sustainable financial products. Green car loans, energy efficiency mortgages, alternative energy venture capital, eco-savings deposits and "green" credit cards, together with social bond, social impact bonds, crowd lending, represent merely a handful of innovative products that are currently offered around the globe. The sustainable character is guarantee by three dimensions: environmental, social and governance ones. However, it is not always easy for a consumer to detach the genuinely sustainable products, given the presence of different opinion about what is meant by "sustainable" or "ethical". Nuclear energy, tobacco or genetically modified organisms are just some examples of "controversial activities", whose financing by means of a sustainable product is currently under discussion. Moreover, some financial products are characterized by attributes that may not be readily assessed or measured, such as products or services that are linked to a charitable donation. For consumers transparency is pivotal. He/she should be able to judge in a clear and easy way whether the sustainable elements of a particular product correspond to his/her requirements.

According to the theoretical frame presented here, a financial product is sustainable to the extent that it is able to create value and not a mere rent and if the value created embeds social and/or environmental aspects. Thus, the capacity of products to answer sustainable standards should be evaluated according to the environmental dimension, the social dimension and the market dimension. However, unlike the concept of sustainability analysed for intermediaries, in this case, it is enough to evaluate the sustainable features of the single product and not its connection with the business model of the bank.

The environmental dimension of the financial products deals mainly with the fragility of ecological systems and their capacity to bear damages and deterioration. In particular, products and services provided by banks meet the environmental standard when the project financed does not pollute the environment. Financial mechanisms are environmental credit risk assessment procedures, initiatives to provide sustainability products and services, support of businesses adopting environment-friendly practices. However, it is widely recognized that the indirect ecological impacts of funding enterprises and projects have to be controlled, managed and followed with attention too: several examples of green product/service innovation, such as the investing in fuel cell companies (utilization of biomass and other renewable energy resources, hydrogen industry, etc.), or the support to ecotourism, can be found in financial sphere.

The social dimension includes initiatives aimed at enhancing welfare (security, health, education), fairly distributed among social classes and genders. Within a territory, the investment in social financing tools encourages the close interaction of stakeholders. There are a number of social initiatives aiming to help the poor, the disabled, the elderly, children and charitable activities. The actions undertaken might create job opportunities for unemployed youth, loans for start-ups and fostering development for women.

The economic sustainability refers to the capacity of the financing products to generate a constant flow of revenues. The products should not only be profitable but also be able to generate positive externalities. Within a territorial system, economic sustainability means the capability, through the most efficient mix of resources, to produce and maintain the highest added value in order to enhance the specificity of territorial products and services and their competitiveness.

While it might be relatively easy to retail financial products with sustainable features, choosing the business model of a financial institution based on sustainable principles is more challenging. This is mainly due to the fact that it is not just a choice of diversifying the supply on the market, but it involves all the operational and governance aspects of the bank. Shareholders might fear a reduction in their revenues and address their funds to more profitable (at least perceived) institutions. In particular, it might be difficult to assess the impact of social and environmental choices, especially for banks whose "production" exploits capital and only marginally labour and natural resources. Hereafter we will first present the market appeal for alternative financial products, such as green bonds. Next, we will focus on the features and the challenges of sustainable banks in the European context.

10.4 Market Appeal of "Alternative" Forms of Finance

New financial products classified as sustainable are available according to region, market and industry structure, and consumer preference: green bonds, social and solidarity financial products are receiving growing attention by the market. In particular, the new remunerative market of green bonds, consisting of green and social finance products, is very appealing, not only for not-for-profit actors but also for profit-driven players. The issuer of these bonds should finance projects (i) aligned with climate bonds taxonomy and (ii) contributing to the reduction of carbon emissions. At international level, insurance systems have been developed for investors that guarantee the development of a transparent market. The Green Bond Principles (GBP), developed by the International Capital Markets Association (ICMA), outline clear requirements for issuers about the definitions of projects to be funded, their selection process, revenue management and reporting. An increasing number of the world's largest banks and corporations have adopted this approach to align their funding to sustainable economy requirements.[4]

The Italian green bond market was created in 2014, with the issuance of a EUR 500 million bond by Hera and a € 3.2 million mini green bond by Enna Energia to finance renewable energy projects. This market has reached a total volume of USD 5.9 billion in mid-January 2018, of which USD 3.3 billion issued in 2017 alone, a value eight times higher than the emissions that were recorded in 2016. Profit-driven actors not usually involved in the sustainable economy (such as private companies and financial organizations) cover about 80% of the volumes. Historically, investments in renewable energy have always dominated all other sectors, but with the growth of the market, some interesting changes have been noted.

[4] For more details on definition of green bonds, see Chap. 2.

In 2017, direct funding for energy efficiency and low-energy buildings more than doubled, compared to 2016, representing 29% of investments. The low-emission transport sector also grew considerably, almost doubling the volume compared to the previous year, thanks to the increase in funding for railway infrastructures and public transport.

The year 2017 has been particularly relevant for the diffusion of the sovereign green bonds. The French Green Bond of EUR 9.7 billion has become the biggest single green bond ever issued. The Fiji Islands announced the issuance of a green state bond of EUR 40 million. Nigeria, in December 2017, issued a EUR 24 million green state bond, the first green state title ever launched by an African country, and the first to obtain certification for the Solar and Land Use Criteria under the Climate Standard Bonds. Sovereign green issues were carried in 2018 in Indonesia, followed by Belgium, Sweden, Morocco and Kenya. United States, China and France dominate the global green bond rankings, accounting together for 56% of the global market in 2017. On a global scale, the green bond market has attracted issuers from 37 different countries, 10 of which made their entrance for the first time in 2017. Table 10.1 reports the green bond issuance in the first four months of 2018.

Table 10.1 Top 15 geographies by issuance of green bond in 2018 (including supranational)

#	*Geography*	*YTD 4/2018 (USD BN)*
1	United States	7.30
2	Belgium	5.55
3	China	4.83
4	France	4.03
5	Supranational	3.72
6	Spain	2.73
7	Sweden	2.27
8	Germany	1.86
9	Indonesia	1.83
10	Italy	1.53
11	Netherlands	1.42
12	Norway	1.24
13	Poland	1.23
14	Canada	1.19
15	Japan	0.94

Source: Adapted from Kaminker and Sachs (2018)

10.5 Sustainable Banks in Europe: Which Challenges?

From a bank's perspective, addressing sustainability implies building both a strategic and a commercial frame of reference. The threats and opportunities resulting for commercial banks that move towards a sustainable model and for ethical and cooperative banks range from risk reduction to profit generation and from business to ideological reasons (Carè 2018). The choice of financial institutions to switch into a sustainable business model have multiple results: they (i) spread a "sustainable business thinking" among their stakeholders, (ii) enhance their reputation and the perceived commitment of the bank, (iii) sustain not-for-profit organizations and projects and (iv) lobby the local and national government to support sustainable projects (according to the bank's capacity of influence). As underlined above, being sustainable is not related to the provision of certain products and services, they come as a consequence. The sustainability involves the offer of an inclusive approach in terms of products, governance, transparency and communication. When reputation is the principal goal of a bank, the effort towards sustainability is not achieved (even though sustainable products are sold), since more convenient tools might replace in the future sustainability to achieve the same goals, that is, better reputation. In this case, we can speak of *weak sustainability* since it is not an internalized choice, but only instrumental. The creation of sustainable value gives a long-term perspective and a stronger commitment to social and environmental dimensions. In this case, we have *strong sustainability*, since the bank switch completely towards a new banking model. In the next paragraphs, the focus is on the sustainability features, either strong or weak, of European banks as classified in Sect. 10.3.1.

10.5.1 Sustainable Profit-Driven European Banks

To highlight the efforts towards sustainability of European banks, we can analyse the top 100 Global Sustainability Index from 2005 to 2019. The companies included in the index are publicly-listed which generate more than USD 1 billion in annual revenue. The ranking is compiled by a Canada-based sustainability-focused financial information company and it is based on 21 key performance indicators (KPIs), covering resource management, employee management, financial management, clean revenue and supplier performance. Among the variables evaluated, there are reduction of carbon emissions and waste, gender diversity in leadership and revenues derived from clean products. In this perspective, sustainability is not only related to the products sold but it involves the whole business model.

The sustainability of profit-driven larger banks might derive from a strategic more than an ethical choice, directed to increase the reputation among shareholders and clients. The analysis can hardly assess whether the bank is strongly or weakly interested in sustainability. However, it may be a starting point for a discussion on these themes. As shown in Fig. 10.1, the number of banks included in the Global 100 has increased since 2005 from 4 to 14. In the case of European banks, their number increases only from 3 to 6. In particular, the gap between the growth trend of European and world banks raises especially after 2009. Not only their number is low but it is difficult to find the same bank constantly ranked in the list. The reasons are manifold but can be related to the fact that these banks put more attentions on reputation than on the creation of sustainable values, and their efforts towards sustainability are only occasional.

To better understand the sustainability commitment, Carè (2018) has analysed the corporate reports published by European banks listed in the Global 100 from 2014 to 2016 using the UN development goals criteria.

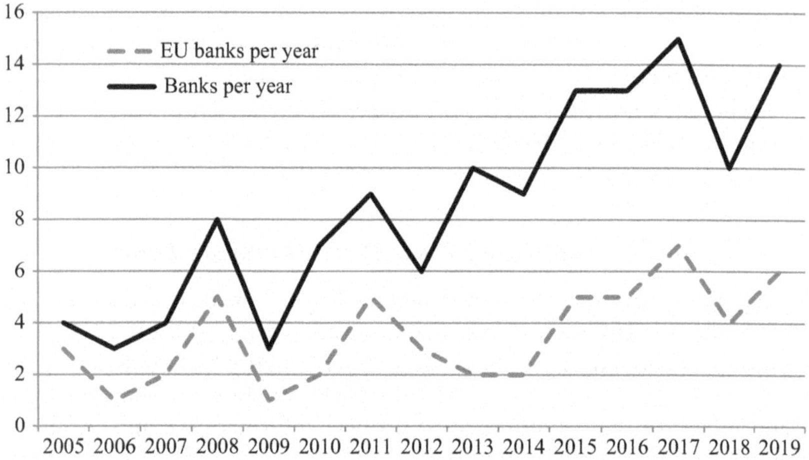

Fig. 10.1 Banks in the global 100 from 2005 to 2019. (Source: Authors' elaboration on "2019 Global 100 Results")

She identifies two main topics on which the commitments to sustainability are based: (i) environmental considerations in terms of direct/indirect impacts and dedicated products and services, and (ii) international engagement and initiatives. The results show how these banks are mainly sensible to enhance climate change actions, together with more economically driven goals such as employment and economic growth. Quality education is relevant for four out of six banks, followed by the creation of the international partnership, and by the goal named "peace, justice and strong institutions". Social issues are less significant: no poverty, zero hunger, reduced inequality, responsible consumption are not popular objectives among sustainable European banks (Table 10.2). It has to be noticed that BNP Paribas and Intesa San Paolo are among the European banks more long-lived in the Global 100. Moreover, BNP Paribas and ING are committed to more than 10 goals out of 17. From the data shown, we can reckon that these banks are more strongly bonded to sustainability, while other banks have a weaker approach.

Table 10.2 Sustainable development goals (SDG) and European banks' approaches

Sustainable development goal	*Danske bank*	*ING*	*BNP Paribas*	*DNB*	*SEB*	*Intesa*	Total
1 No poverty			√				1
2 Zero hunger			√				1
3 Good health and wellbeing	√		√				2
4 Quality education	√	√	√	√			4
5 Gender equality	√		√				2
6 Clean water and sanitation		√	√				2
7 Affordable and clean energy		√	√			√	3
8 Work and economic growth	√	√	√	√	√	√	6
9 Industry, innovation and infrastructure		√			√		2
10 Reduced inequality			√				1
11 Sustainable cities and communities		√	√				2
12 Responsible consumption		√					1
13 Climate action	√	√	√	√	√		5
14 Life below water			√				1
15 Life on land			√				1
16 Peace, justice and strong institutions		√	√		√		3
17 Partnership for the goals	√	√	√			√	4
Total	6	10	15	3	4	3	41

Source: Authors' elaboration on data from Carè (2018)

Two main conclusions can be drawn from the above analysis. First, larger banks committed to sustainable issues prefer a more secure and easy-to-sell involvement in environmental goals compared to social issues, which involve a precise (political) choice from the financial institutions. Second, maintaining a strong commitment to these objectives requires constancy and efforts and does not repay the investments made in the short term in terms both of revenues and of reputation. The role of profit-driven banks is relevant in the European context and their effort to include some sustainable aspects in their business model is precious. The challenge is to switch from an instrumental use of sustainability to a more tangible involvement in supporting a socially and environmentally sensible creation of value.

10.5.2 Not-for-Profit and Specialized European Banks

Even though less relevant in terms of market shares, banks with a not-for-profit aim are pivotal in supporting sustainable finance. Even if usually jointly considered, not-for-profit and specialized banks have different approaches to sustainability. As mentioned in Sect. 10.3, specialized banks do not necessarily have a cooperative nature and they are completely devoted to environmental and social initiatives. We consider not-for-profit banks financial intermediaries with a cooperative form but different from specialized banks since in this classification cooperative banks might be more or less perceptive to these themes according to their priorities set.

Specialized Banks

For over 30 years, alternative financial institutions have been created in order to disseminate ethical and solidarity-based financial models in the European economic and political area. A specialized bank, such as an ethical bank, guarantees that the administered deposits will be channelled to cultural, social and environmental projects. Through their activity, they promote social inclusion, sustainable development, development of social economy and social entrepreneurship. Ethical banks also help to raise public awareness about the role of money and the failure of the economy based on short-term approaches and profit as the only objective.

As other local banks, an ethical bank is rooted in the territory in which it operates and exploits its socio-economic networks. This allows to have full knowledge of its clients and their projects. A fundamental value for them is transparency, especially towards customers, both in the origin and use of money and in credit and business management. The European

Table 10.3 Comparison between European *ethical* and *systemic* bank (percentage)

		2016	*2011*	*2006*
Loans/total assets	European ethical banks	73.42	75.25	64.87
	European systemic banks	38.53	34.62	32.93
Deposits/total assets	European ethical banks	80.87	69.10	62.31
	European systemic banks	42.15	32.57	33.98
Net equity/total assets	European ethical banks	11.22	11.22	10.94
	European systemic banks	5.63	4.39	3.86

Source: Authors' elaboration on Cavallito et al. (2017)

Federation of ethical banks counts 26 members (12 banks, 8 savings and loan cooperatives, 4 investment companies and 2 foundations) across 15 European countries, with different sizes and legal forms.

A specialized bank can play a significant role in the development of a new banking system in Europe. Their business model, mainly based on the intermediation between borrowers and lenders, can foster a return to the root of banking, closer to the real economy than to the financial speculation. This emerges from data in Table 10.3, comparing the percentage of loans on total assets. Moreover, ethical banks rely more on client deposits to fund their activities unlike systemic banks that issue bonds or use deposits from second banking market. Finally, the ratio between net equity and total assets underlines the strong capital position of ethical banks; in other words, they are closer to the OTH (originate to hold) model, than to the OTD (originate to distribute) one.

Specialized banks in general, and ethical banks, in particular, can represent an opportunity for the development of a strong sustainable business model in Europe. Their attention to offering services to the real economy, their reduced volatility on the market and their stronger capital position underlines the care to the old-style banking method, that is, the intermediation between demand and supply. The value created is conveyed to the productive economy involved in social and environmental projects. The exclusiveness might reduce their capacity to differentiate the risks; however, it strengths the linkages with their customers, sensitive to these themes.

Cooperative Banks

The role of cooperative banks in sustainable finance has been recognized both in the Commission's High Level Group Report on Sustainable finance and in the European Parliament's own initiative report, where it is stated:

"we should also acknowledge the leading role played by cooperative and community finance in pioneering green investments" (EACB 2019).

Cooperative banks can exploit a large network in Europe with 58,000 branches and 209 million clients and they can play a key role in financing socially and environmentally sustainable projects. Thanks to their decentralized model, these banks are crucial to integrate productive activities with a sustainable use of capital. As retail banks, they channel finance to the real economy and to local SMEs, key actors in job creation. Moreover, by statute they reinvest significant quotas of their available profits in the society they belong, by supporting social and cultural projects with a long-term perspective, and encourage their clients/members to follow this behaviour.[5] Their success depends strongly on the economic, ecological and social wellbeing of their operating territory. To enhance sustainable finance, the stability of fundamental parameters such as environmental policy, taxation, prudential requirements is essential. In this sense, cooperative banks have carried out sustainable finance since their birth, placing deposits for the benefit of the local real economy. However, the complexity and the continuous review of the regulatory framework is affecting the capacity of cooperative banks to finance real economy (e.g. the long-term funding), and reducing their peculiarities as a result of a standardization process.

The threat for cooperative banks is that, not being specialized banks, they are likely to follow a weak sustainable strategy. In some countries, cooperative banks are key player in green bonds issuing. In Germany, DZ BANK has been active in this market segment since 2013 and it is one of the ten leading syndicate banks. The fact that sustainability is a major aspect of the cooperative principle and culture is proving to be an advantage. Beside DZ Bank in Germany, also Crédit Agricole in France and Rabobank in the Netherland are in the list of market leaders. The question is whether cooperative banks have entered this market for ethical or for profitable reasons.

If we compare the green bond market with energy efficient mortgage loans, we might better assess the sustainability of banks. The stock of buildings in the EU is relatively old and, therefore, more energy consuming: renewing it is crucial if the energy consumption is to be reduced. Renovation leads, among other, to higher property values, lower energy

[5] For more information on future challenges and perspective of cooperative banks, see Migliorelli (2018).

bills, hedging of energy price with particular relevance in peri-urban and rural households. Dedicated loans are a critical financial instrument to redirect private capital into energy efficiency investments and lowering carbon emissions. However, this market is not as attractive as the green bonds.

There are, however, signs of the commitment of cooperative banks to investments possible less remunerative than the green bonds, but relevant in terms of energy savings. In Spain, Grupo Cooperativo Cajamar has financed in 2016 about 5.000 transactions for a total amount of EUR 600 million. In Austria, the regional banks of the RBI have contributed EUR 411 million. In the Netherlands, Rabobank has contributed EUR 50 million in 2016. In France, Crédit Mutuel has financed 5.400 projects for a total amount of EUR 100 million. Crédit Agricole's regional banks have financed home energy renovations in 2016; over 104,000 offers have been made totalling over EUR 2.1 billion. German cooperative banks had a high market share (more than 30%) in offering promotional housing loans of KfW regarding energy efficiency in 2016. Groupe BPCE was the first pilot bank to sign an agreement with the European Commission in 2012 to organize the financing of the energy transition in the territories. Households have benefited from EUR 2.8 million in loans from participating Banques Populaires and Caisses d'Epargne, allowing a total final investment of EUR 29.9 million and an energy saving of 56.68 GWh/year (EACB 2017).

Cooperative banks might play an active role by promoting within their network the distribution of services and investment or savings products in favour of sustainable development. Their expertise in gathering local needs and in supporting actors should be exploited to address stakeholders to sustainable productive initiatives. These banks should put in place innovative solutions to use in a responsible manner the resources that are indispensable to their activities. However, cooperative banks must not be lured by short-term profits coming from green financial markets: they must support socially and environmentally sustainable economic growth even at the cost of renouncing to immediate high performances.

10.6 Conclusions and Further Perspectives

Moving from the definition of sustainable finance in terms of value, the chapter attempts to read the recent phenomenon of the growth of the ESG—that is, environment, social and governance—investments in the banking sector. The scrutiny of the different forms of bank present in the European market, according to their involvement with sustainable products, has allowed defining a weak versus a strong form of sustainability. While it is easy and appealing to sell green products in the actual economic context, this is not sufficient to match the sustainable policy as defined in this chapter, that is, it is not yet the creation of genuinely sustainable value, since still moved by redistributive motivations. In this case, sustainability might be used to enhance the reputation of the financial institutions (consider the case of blockchain offering financial green products). A strong effort towards sustainability involves all the aspects of the banking model. In this perspective, the life horizon is the long term and the switch to a sustainable business frame should not be considered as an ephemeral fashion, but as a new banking paradigm. The differences among institutions in term of ownership may still play a role, since listed banks should pay returns to their shareholders. However, the emphasis should not be on the rent, but on the added value generated by investments. In this context, cooperatives and specialized banks can be less restricted in addressing their investment choices.

Sustainable finance is still in its infancy. Actions should be taken to increase its relevance in the economy. A first step could be the development of a common taxonomy for sustainable assets, with minimum standards. Second, a stable legislative and regulatory framework, able to catch the peculiarities of each bank models and to enforce an ecosystem able to cater to different needs and longer-term approaches, is pivotal. However, key parameters for environmental policy, taxation and prudential requirements should be clearly defined in order to favour strong sustainable banking model. Finally, a particular effort should be put in reviewing the methods of computation of GDP, so that value (created also by the financial intermediaries) increases the total, while negative externalities reduce it.

References

'2019 Global 100 Results'. (2019, January 22). *Corporate knights.* https://www.corporateknights.com/reports/2019-global-100/2019-global-100-results-15481153/

Carè, R. (2018). *"Sustainable banking: Issues and challenges", Palgrave Pivot.* Cham: Palgrave Macmillan.

Cavallito, M., Isonio, E., & Meggiolaro, M. (2017). *Ethical and sustainable finance in Europe.* Firenze: Finanzaetica.

Dombret, A. (2018). *"Greener finance – Better finance? How green should the financial World be?" Bundesbank symposium 'Banking supervision in dialogue'*. Frankfurt/Main: Deutsche Bundesbank.

EACB. (2017). *Towards a sustainable finance framework in EU: The views of Co-Operative Banks – Policy paper*. European Association of Co-operative Banks.

EACB. (2019). *Annual report 2018*. European Association of Co-operative Banks.

Goglio, S. (2002). *Fattori immateriali e sviluppo. Economia e società regionale*. Milano: FrancoAngeli Editore.

Kaminker, C. R., & Sachs, S. (2018). *The green bond*. SEB. https://www.eticanews.it/wp-content/uploads/2018/05/SEB-The-Green-Bond-May-2018-Final-media.pdf

Marshall, A. (1920). *Principles of economics*. London: Macmillan.

Mazzucato, M. (2018). *The value of everything: Making and taking in the global economy*. London: Allen Lane, an imprint of Penguin Books.

Migliorelli, M. (Ed.). (2018). *New cooperative banking in Europe. Strategies for adapting the business model post-crisis*. London: Palgrave Macmillan.

Silver, N. (2017). *Finance, society and sustainability – How to make the financial system work for the economy, people and planet*. London: Palgrave Macmillan.

Sustainable Saving and Investing. (2019). *Financial products with a particular focus on people, the environment and sustainable business*. http://www.sustainablesavingandinvesting.be/.

UN Environment Inquiry. (2016). *Financing the future: Report of the Italian national dialogue on sustainable finance*. Geneve.

CHAPTER 11

Green Finance Today: Summary and Concluding Remarks

Marco Migliorelli and Philippe Dessertine

The contents included in this chapter do not necessarily reflect the official opinion of the European Commission. Responsibility for the information and views expressed lies entirely with the authors.

M. Migliorelli
IAE Université Paris 1 Panthéon-Sorbonne (Sorbonne Business School), Paris, France

European Commission, Brussels, Belgium

P. Dessertine (✉)
IAE Université Paris 1 Panthéon-Sorbonne (Sorbonne Business School), Paris, France
e-mail: philippe.dessertine@univ-paris1.fr

M. Migliorelli, P. Dessertine (eds.), *The Rise of Green Finance in Europe*, Palgrave Studies in Impact Finance,
https://doi.org/10.1007/978-3-030-22510-0_11

11.1 The Role of Finance in Fostering a Sustainable Economy

The adoption of the Sustainable Development Goals (SDG)[1] in September 2015 and the Paris Agreement[2] reached in December of the same year in the United Nations Framework Convention on Climate Change (UNFCCC)[3] landmarked the commitment of the international community to the sustainability of human activities and to the fight against climate change. Since then, even if in an increasingly variable political environment,[4] policy makers and scientists focused on the strategies to put in place in order to deliver on the most ambitious objectives and targets. In this respect, the role of finance in supporting the transition has been continuously underlined by both governments and international organisations (e.g. EC 2018a).

In such a context, *green finance* has been progressively recognised as a decisive enabling factor in an attempt to reach the environment-related goals. Climate change mitigation, climate change adaptation and preservation of soil and water are amongst the activities that will require a continuous and increasing flow of dedicated financial resources in the next decades. Even if it has a relatively short history (the first *green bond* was issued by the European Investment Bank only in 2007), *green finance* can already be considered one of the most promising components of the sustainable finance landscape. Featured by multiple-digits growth since its inception, the green

[1] The 17 Sustainable Development Goals were adopted by world leaders in September 2015 at an historic UN Summit in New York and officially entered into force on 1 January 2016. They are the key part of the 2030 Agenda for Sustainable Development and they universally apply to all countries, hence being a mobilising effort to end all forms of poverty, fight inequalities and tackle climate change.

[2] The Paris Agreement's central aim is to strengthen the global response to the threat of climate change by keeping the global temperature rise this century well below 2 degrees Celsius above pre-industrial levels and to pursue efforts to limit the temperature increase even further to 1.5 degrees Celsius. Additionally, the agreement aims to strengthen the ability of countries to deal with the impacts of climate change. The Paris Agreement requires all Parties to put forward their best efforts through nationally determined contributions (NDCs). This includes requirements that all Parties report regularly on their emissions and on their implementation efforts. The Paris Agreement entered into force on 4 November 2016, 30 days after the date on which at least 55 Parties to the UNFCCC accounting in total for at least an estimated 55% of the total global greenhouse gas emissions have deposited their instruments of ratification, acceptance, approval or accession.

[3] The Convention that brought the Paris Agreement is also called COP21, as indicating the 21st Convention of the Parties.

[4] In June 2017, US President Donald Trump announced his intention to withdraw his country from the Paris Agreement. Under the agreement itself, the earliest effective date of withdrawal for the United States is November 2020.

finance market has reached in 2017 a record level of global issuance, with new *green bonds* for USD 155.5 billion.[5] Nevertheless, this is considered by many as a *drop in the ocean* when compared with the effective needs for assuring an environmentally neutral and sustainable economy.[6]

It can be argued that the growth of *green finance* observed so far has been the result of two intertwined elements: the active engagement of international financial organisations and the emergence trough time of specific market incentives for private investors to issue or buy green securities. International financial organisations (in particular, the European Investment Bank and the World Bank) have triggered the inception phase of the *green finance* market by issuing the first securities and still today remain the largest issuers of *green bonds*. For these organisations, embracing *green finance* is a concrete way to align their institutional mission to their funding arrangements. As concerns the engagement of private investors, the observed emergence of specific market incentives can be linked in particular to reputational gains, corporate social responsibility acknowledgement (by existing and perspective clients) and the realisation of specific business opportunities. As also observed in literature (e.g. Friede et al. 2015; Anselmsson et al. 2014), for a company to be recognised as engaged in sustainable activities (e.g. by carrying out environmental projects financed by green securities) can bring concrete benefits in terms of customer satisfaction, customer retention and market positioning.[7] The strong demand of green-labelled securities from institutional investors can be also seen in this light. On the other hand, *green finance* has ascended as a concrete source of additional revenues for funds managers, stock exchanges and green certification bodies, which hence developed new products and services, in this way significantly supporting the nascent market. In addition, the incentive of embracing *green finance* for issuers should be further assessed as following an emerging strand of literature aiming at demonstrating that green securities could incorporate a not-negligible market premium coming from investors' preferences for environmentally friendly investments (e.g. Zerbib 2019). Nevertheless, wider evidence is still needed to fully support this conclusion.

[5] Source: Climate Bond Initiative (2018).

[6] As an example, investments of around EUR 520–575 billion annually have been estimated to be necessary in the EU only in order to achieve a net-zero greenhouse gas economy in the 2050 horizon. Source: EC (2018a).

[7] In the financial market, the recognition as engaged in sustainable activities is particularly important for dual bottom-line institutions, such as cooperative or saving banks (see also Migliorelli 2018).

11.2 Market Forces Alone Will Not Be Sufficient to Mainstream *Green Finance*

Can it be assumed that market forces alone will be sufficient to trigger the needed amount of resources to foster an environmentally sustainable economy? The answer to this question is straightforward: no. As a matter of fact, a number of substantial limits can be observed when analysing the key current market dynamics featuring the *green finance* sector under such a point of view.

First, both demand and supply of green securities should be considered to be still in an early stage of development. In this respect, it can be argued that the *green finance* sector lacks some of the necessary enabling factors that could allow buyers and sellers of green securities to easily enter the market. On the one side, existing and latent demand is still poorly channelled towards green securities due to the absence of easy-to-understand references. This is the case in particular of widespread green labelling criteria and labels (with the relevant exception of *green bonds*) and the possibility for retail investors to easily transmit their investment preferences for green placements to the intermediaries that manage their funds. On the other side, it should be observed that the supply of green securities is today very limited, as concerns both size and products, in this way representing a concrete cap to the market potential. In this regard, the not-negligible costs of screening, labelling, disclosure and control-related activities to be borne by potential issuers factually represent a significant hurdle to the smooth development of the market. These costs can also explain why the issuance of green securities has been so far restricted to large companies in the private sector.

Second, and more important, it can be argued that market forces alone will be unlikely sufficient to trigger an appropriate amount of resources to foster an environmentally sustainable economy as today little incentive exists for market investors to *fully* factor-in environmental risks in their decision-making processes (which include the choice of type of investments to undertake and the way they are financed). In fact, many of these risks materialise only in the long term (e.g. in the form of higher frequency and magnitude of climate-related extreme weather events, such as droughts, floods and storms), and they mainly result in externalities for which the original polluters do not directly pay. This seems true also when advanced frameworks aimed at imposing a direct cost to the largest polluters have been put in place.[8]

[8] An example is the European Union (EU) emissions trading system (ETS). Launched in 2005, the ETS is a tool that would support the effort to cut greenhouse gas (GHG) emis-

11.3 The Efforts Needed to Mainstream *Green Finance*

The development of *green finance* has to be seen as an instrument and not as a mere end. In this respect, mainstreaming *green finance* can indeed play a significant enabling role to reach all the environmental goals. To do that, it can be argued that the accomplishment of three principal conditions is necessary: environmental risks are properly included in the investors' decision-making processes, market demand is effectively channelled towards green investments and *additionality* is adequately encouraged by policy makers.

One of the most significant hurdles to the systematic inclusion of environmental risks in the investors' decision-making processes results from the lack of reliable data and research on the economic consequences of the environmental risks. Even if the impact of human activities on climate change is progressively getting accepted (e.g. IPCC 2018), the specific link between environmental risks and economic and financial risks still needs a considerable amount of work before it can be considered as completely mastered or even understood. In particular, the specific channels and the magnitude of the transmission of environmental risks on the real and financial economy today still remain largely unidentified. As a matter of fact, such lack of information represents an impediment to the (technical and political) capacity of policy makers to shape effective adjustment mechanisms and policies in order to make the *polluter pays* principle completely effective. In addition, an incorrect management of environmental risks due to the lack of reliable data can have important (and potentially systemic) negative effects on some sectors of the financial market that are indirectly exposed to these risks. This is the case in particular of the insurance and banking industries. In fact, these industries could progressively suffer from increasingly high unexpected losses due to higher payments on the insured risks or to higher impairments on the loans issued, as a consequence of a fall

sions by at least 40% by 2030 compared to 1990 levels. It obliges more than 11,000 power plants and factories to hold (and pay) a permit for each tonne of CO_2 they emit in order to provide a financial incentive to pollute less. Companies have to buy them through auctions and the price is affected by demand and supply (even though some of the permits are allocated for free, particularly in sectors at risk of having companies to move production to other parts of the world with laxer emission constraints). One of the main weaknesses of such a system in fighting climate changes lies in the global dimension of the problem. As a matter of fact, global warming can be hardly managed without a global commitment. For such a reason, it could be argued that climate change will keep impacting also the virtuous countries if a global solution is not reached.

in the value of the assets following a higher occurrence of climate-related extreme weather events. Hence, an argument should be made according to the idea that a thorough analysis of the relation linking environmental risks and economic and financial risks has to be one of the key areas of action for policy makers (and academicians) in the next few years.

As concerns the proper channelling of the demand, it can be easily argued that a better steering of all the types of investors towards green securities would greatly encourage the development of *green finance* by triggering a further market-based endogenous growth. To do that, a number of actions could be put in place, both by the financial industry and by policy makers. These actions can include in particular the definition of an universal taxonomy of green activities (which could give certainty on which investments are to be considered "green"), specific labelling criteria for green financial products, disclosure requirements in public corporate reporting and the incorporation of the clients' preferences with respect to green finance in the counselling activities of financial intermediaries. As a matter of fact, in some policy contexts, many of these initiatives have been already launched, as in the European Union (EU),[9] and will be gradually phased-in.

Nevertheless, as of today, it can be expected that a full inclusion of environmental risks in the investors' decision-making processes and a complete channelling of demand towards green securities will hardly materialise in the short term. In addition, any possible action to be launched to these ends may eventually present a certain degree of ineffectiveness, due in particular to the highly innovative profile of the subject and the absence of *silver bullet* in terms of technical solutions available. A situation of *environmental market failure* or of suboptimal investment levels can indeed still materialise also in the most advanced policy frameworks. For these reasons, a third condition for directing an adequate level of financial resources towards the investments necessary to reach the environmental goals is that *additionality*[10] is adequately encouraged by policy makers when needed. To this extent, among other possibilities, policy makers can play a role in particular by implementing specialised financing facilities aimed at financing environmental investments. These financing facilities could be backed by lending, guarantees or equity instruments (or a combination of the three), in which government spending could eventually cover part of the risk, including when necessary the so-

[9] For a more detailed analysis, see Chap. 6.

[10] By the term *additionality* it is here intended a policy-triggered additional amount of financial resources.

called *first loss*. Furthermore, policy makers could endorse or lead a wide set of initiatives aiming at encouraging the participation of market investors to environmental investments (e.g. by sponsoring a pipeline of environmental investments or promoting specific public-private partnerships).

Finally, it should be stated that, to mainstream *green finance*, a specific role should be played by the banking sector. In this respect, the relevance of banks is given by the high level of funds intermediated[11] and by their capacity to effectively reach small and medium enterprises (SMEs) and households, which are often crucial actors for the success of environmental initiatives. As the three main conditions to mainstream *green finance* (environmental risks are properly included in the investors' decision-making processes, market demand is effectively channelled towards green investments and *additionality* is adequately encouraged) ultimately converge in banking operations, it can be argued that specific actions should be put in place by policy makers to specifically induce banks to embrace *green finance*. These actions can span from prescriptive measures (e.g. the imposition of disclosure and reporting requirements having as object the banks' sustainability activity) to provisions encouraging the lending to environmentally conscious firms (e.g. by easing some regulatory requirements for loans directed to environmental investments), to the promotion of instruments able to calibrate the pricing of financial products on the basis of the effective environmental risks of the borrowers (e.g. by encouraging the development of methodologies of rating factoring-in these risks).

11.4 *Green Finance* Beyond Finance

Green finance still has a long way ahead before it can be considered a stable component of the modern financial landscape. Besides market-based dynamics and the possible policy actions that can be put in place to further develop the sector, it is increasingly evident that the fortune of *green finance* will decisively depend on the level of political commitment of the international community towards the environmental goals. In this respect, the EU has taken the lead and has already tabled a number of concrete initiatives. Among the most relevant ones are the issuance of the European Commission's *Action Plan for a greener and cleaner economy*,[12] the finalisation of some specific legislative proposals to develop sustainable finance[13]

[11] In Europe, the total assets of the banking sector peaks up to 350% of the aggregate GDP. Source: EBF (2012).

[12] EC (2018b). For a wider description, see also Chap. 6.

[13] Following the *Action Plan for a greener and cleaner economy*, in May 2018 the European Commission adopted a package of implementing measures including: a proposal for a regula-

and the release of the *European strategic long-term vision for a prosperous, modern, competitive and climate neutral economy*.[14] Nevertheless, the EU is responsible for only 10% of global greenhouse gas emissions.[15] In point of fact, only a global governance for the fight to climate change can produce the shift in paradigm needed for reaching environmental goals. Realising the objective of limiting the rise of global temperature this century well below 2 degrees Celsius above pre-industrial levels and pursuing efforts to limit the temperature increase even further to 1.5 degrees is only possible with the involvement of all the major stakeholders, including the most polluting industries, municipalities and the civil society. To this extent, intermittent political commitment can indeed risk to dilute the efforts and the results reached so far. On the contrary, a convergence of intentions (and interests) would produce the necessary *signalling effect* that would powerfully stimulate the development of all the concerned industries, including the nascent *green finance* industry.

References

Anselmsson, J., Bondesson, N. V., & Johansson, U. (2014). Brand image and customers' willingness to pay a price premium for food brands. *Journal of Product and Brand Management, 23*(2), 90–102.

Climate Bonds Initiative. (2018, January). *Green bonds highlights 2017*. London.

EBF. (2012). *EU's banking sector: The world's largest banking system. Facts and figures 2011/2012*. Brussels.

European Commission (EC). (2018a). *A clean planet for all. A European strategic long-term vision for a prosperous, modern, competitive and climate neutral economy*, COM(2018) 773 final. Brussels.

tion on the establishment of a framework to facilitate sustainable investment (this regulation establishes the conditions and the framework to gradually create a unified classification system—or taxonomy—on what can be considered an environmentally sustainable economic activity); a proposal for a regulation on disclosures relating to sustainable investments and sustainability risks and amending Directive (EU)2016/2341 (this regulation aims at introducing disclosure obligations on how institutional investors and asset managers integrate environmental, social and governance factors in their risk processes); a proposal for a regulation amending the benchmark regulation (to create a new category of benchmarks comprising low-carbon and positive carbon impact benchmarks, which should provide investors with better information on the carbon footprint of their investments). See also Chap. 6.

[14] EC (2018a).

[15] Source: World Research Institute. Data referred to 2014. As concerns the other most polluting countries, China counts for about 26% of the GHG emissions, the U.S. for 15%, India for 6%, Russia for 5%, Japan for 3%, Brazil for 2%.

European Commission (EC). (2018b). *Action plan: Financing sustainable growth*, COM(2018) 97 final. Brussels.

Friede, G., Busch, T., & Bassen, A. (2015). ESG and financial performance: Aggregated evidence from more than 2000 empirical studies. *Journal of Sustainable Finance and Investment, 5*(4), 210–233.

Intergovernmental Panel on Climate Change (IPCC). (2018). *Special report. Global warning of 1.5 °C*. Geneva.

Migliorelli, M. (Ed.). (2018). *New cooperative banking in Europe. Strategies for adapting the business model post-crisis*. London: Palgrave Macmillan.

Zerbib, D. O. (2019). The effect of pro-environmental preferences on bond prices: Evidence from green bonds. *Journal of Banking and Finance, 98*, 39–60.

Index[1]

[1] Note: Page numbers followed by 'n' refer to notes.

M. Migliorelli, P. Dessertine (eds.), *The Rise of Green Finance in Europe*, Palgrave Studies in Impact Finance,
https://doi.org/10.1007/978-3-030-22510-0

The manufacturer's authorised representative in the EU is Springer Nature Customer Service Centre GmbH, Europaplatz 3, 69115 Heidelberg, Germany. If you have any concerns regarding our products, please contact ProductSafety@springernature.com

Printed and bound by CPI Group (UK) Ltd, Croydon, CR0 4YY

30/09/2026

02225649-0007